Julian Cribb is a distinguished science wr· ‧ than thirty awards for journalism. He was ıder of the influential *ScienceAlert* ks, including *The Coming Famin*

★

Poisoned Planet is a crucially important book, dealing with an environmental threat that could be as serious (or even more serious) than climate disruption—the toxification of Earth. If climate disruption gets totally out of control there is a crazy solution—geoengineering—that will likely be tried. For toxification we have no last ditch efforts even under discussion.

If you care about your children, read this book.

—**Paul R. Ehrlich**, Bing Professor of Population Studies, Stanford University

Most of us are already aware that humanly-made chemicals pervade our environment and threaten our health. *Poisoned Planet* documents the terrifying extent of chemical pollution, the lack of testing and regulation, and the survival threat to countless other species and whole ecosystems. Despite the banning of a few chemicals in at least some countries, the overall problem has actually worsened dramatically since Rachel Carson's day.

We need a *Silent Spring* for the twenty-first century, and Julian Cribb has given us one.

—**Richard Heinberg**, Senior Fellow, Post Carbon Institute, and author of *The End of Growth*

'Shines a much-needed floodlight on one of the world's most urgent—and underappreciated—environmental problems'

—Professor Ravi Naidu, Cooperative Research Centre for Contamination Assessment and Remediation of the Environment

'Deserves to be widely read—not just by environmental activists, but by anyone who cares about their health and the wellbeing of their children.'

—Ian Lowe, Emeritus Professor of Science, Technology and Society, Griffith University

'Casts a penetrating light into the cloud of information (and disinformation) surrounding modern chemicals and makes a compelling case for caution and prevention... *Poisoned Planet* offers ways you can help clean up the Earth and defend the "human right not to be poisoned".'

—Wellbeing Magazine

'Cribb's message—emphasising the need for both information and precaution—warrants our serious attention. The problem of persistent environmental chemical contamination could well have long-lasting effects on human biology, health and longevity, and we owe it to ourselves and coming generations to debate the questions he raises.'

—Professor Tony McMichael, *The Conversation*

'According to Julian Cribb, something even larger than climate change is happening—the poisoning of the planet... It's a human problem, with human solutions.'

—Sydney Morning Herald

POISONED PLANET

HOW CONSTANT EXPOSURE TO MAN-MADE CHEMICALS IS PUTTING YOUR LIFE AT RISK

JULIAN CRIBB

First published in 2014

Allen & Unwin
83 Alexander Street
Crows Nest NSW 2065
Australia
Phone: (61 2) 8425 0100
Email: info@allenandunwin.com
Web: www.allenandunwin.com

Cataloguing-in-Publication details are available
from the National Library of Australia
www.trove.nla.gov.au

ISBN 978 1 76011 046 8

Internal design by Alissa Dinallo
May by Map Graphics, Brisbane
Index by Puddingburn
Set in 11.5/15.75 Bembo by Bookhouse, Sydney
Printed and bound in Australia by Griffin Press

10 9 8 7 6 5 4

This book is dedicated to Professor Ravi Naidu
for his unfailing insight, wisdom, inspiration
and example in addressing
one of the world's greatest challenges

CONTENTS

LIST OF TABLES AND FIGURES

TABLES

FIGURES

PREFACE

This book is a wake-up call.

It deals with an existential threat to all humanity and, potentially, the whole of life on Earth. One that is entirely of our own making, and which we alone can dispel.

Most people know that some chemicals are not good for us and there are too many of these things in our food and environment. There are many 'small picture' stories about chemical accidents, polluted sites and chemicals in food and consumer products.

These stories are mere pixels in a very much larger global image, one that now directly concerns and affects every person living on the planet. This book explores the big picture, bringing together peer-reviewed science and evidence from trustworthy international sources. It is about a subject larger even than climate change, one that will affect all of us for centuries to come.

It is written in plain, non-technical language for the ordinary citizen, woman or man, who wants to understand the real impact of their actions as a consumer on themselves, their children and on future generations—and who wants to do something about it.

The subject matter makes for uncomfortable reading, and for that reason the books is short and factual, with plenty of notes for those who wish to verify its claims or investigate its sources. Although the topic is grim, this is very much an optimistic book. Whatever we have done to our world, we can undo, and in ways that will lead to better health, greater prosperity and more opportunity. But only through an act of willing cooperation by humans across the world—perhaps the first such truly global act by *Homo sapiens,* and the one that ultimately defines us.

Read on and be not daunted. Rather, be encouraged to help build a better, cleaner world.

Julian Cribb

CHAPTER 1
CHEMICAL COLOSSUS

To cause awareness is our only strength.

W. Eugene Smith, *Minamata*, 1975

The pallid light of a mid-winter afternoon, filtering through a tiny window set high in the wall of the small bathroom, illuminated mother and child in a moment of exquisite tenderness and pathos. Eugene Smith shifted uncomfortably in the cramped chamber to reframe the image: shrapnel wounds sustained in Okinawa as a war correspondent almost thirty years earlier still troubled him. Sightless, deaf, lame, claw-handed and emaciated, Tomoko Uemura lay helplessly in the bath, cradled in her mother's loving arms. Sixteen years earlier she had sustained terrible damage as she still lay in the womb, the venom that crippled her leaching unseen and undetected from the outlet pipe of the nearby chemical plant into the surrounding sea that furnished the food for her village.

Smith was a veteran photographer and photo-journalist who had seen it all—war, suffering, human courage, character and compassion, industry and politics—and depicted it in an epic series of photographic essays, many published in *Life* magazine over several decades. Aroused by growing evidence of the devastation being inflicted on ordinary people by chemical pollution, Smith and his wife Aileen moved in 1971 to the town of Minamata, Japan, following reports of a mysterious disease that had been afflicting its inhabitants since the mid-1950s, to document its impact in images and words. The disease was caused by methyl mercury, a substance so poisonous it has no 'safe' level of exposure, no matter how small the dose. It originated in discharges from the local chemical plant. He wrote:

> The nervous system begins to degenerate, to atrophy. First, a tingling and growing numbness of limbs, and lips. Motor functions may become severely disturbed, the speech slurred, the field of vision constricted. In early, extreme, cases victims lapsed into unconsciousness, involuntary movements and often uncontrolled shouting. Autopsies show the brain becomes spongelike as cells are eaten away. It is proven that mercury can penetrate the placenta to reach the fetus, even in apparently healthy mothers.[1]

The Smiths came for three months. They stayed three years and it almost cost the photographer his life. On 7 January 1972, barely a month after he captured the immortal image of Tomoko and her mother—later to be known as the 'Madonna of Minamata'—he accompanied a group of mercury-poisoning victims to cover a meeting arranged with a manager of the Chisso company which was responsible for running the chemical plant, and thus also for the mercury-laden discharges into local waters where they contaminated the marine food chain on which locals relied. The manager failed to show up. Smith later recounted:

> Suddenly, a mob of workers rounded a factory building . . . They hit. They hit me hardest, among the first . . . The last exposure, bad, blurred, shows the man on the left, his foot at that moment finishing with my groin, reaching my cameras. The man on the right was aiming for my stomach. Then four men raked me across an upturned chair and thrust me into the hands of six who lifted me and slammed my head against the concrete outside, the way you would kill a rattlesnake if you had him by the tail.[2]

Battered and bruised, his cameras smashed, Smith survived but lost partial sight in one eye. It turned out to be his last assignment and he died in 1978.[3]

The bludgeoning of Eugene Smith exemplifies the lengths to which some organisations and individuals will go in order to stall growing awareness of the effects of toxins discharged by their enterprises on the community. In spite of such attempts to silence the truth, awareness has slowly spread, more so in some societies than others; more in some social strata than others. But the warning has spread neither far nor fast enough: today, most people still have barely an inkling of the universal chemical inundation to which they are now subject, daily, and of the growing peril which we—and all our descendants—face. If the dawn of that awareness for the educated publics of North America and Europe came with the publication of Rachel Carson's devastating book *Silent Spring* in 1962, where she revealed the impact of certain pesticides used in the food chain on wildlife and humans, then

Eugene Smith's searing image of the 'Madonna of Minamata', transcending words and languages, was the shot heard round the world.

The subject matter of this book is plain, unvarnished science, as brutal in its way as the fists and boots that fell on Eugene Smith. But it is the truth, insofar as any system devised by humans is able to determine and describe such things.

* * *

Earth and all life on it are being saturated with man-made chemicals in an event unlike anything which has occurred ever before, in all four billion years of our planet's story. At almost every moment of our lives, from conception to death, we are exposed to thousands of man-made substances, some known to be deadly in even minute doses and most of them unknown in their effects upon our health and wellbeing or upon the natural world. These substances enter our bodies with each breath, each meal or drink, the things we touch or meet in our journey through each day. There is no escape from them.

Ours is now a poisoned planet, its whole system infused with the substances we deliberately or inadvertently produce in the course of extracting, making, using, burning or discarding the many marvellous products on which our modern life depends. Relative to the span of human history, this has all happened quite quickly and has grown so rapidly that most people are still unaware of the extent or scale of the peril in which it places each of us and our posterity. Our present plight has crept up on us unseen, piecemeal, with infinite subtlety and frequent inadvertence, in a social climate of trusting acceptance, over barely the span of a single human life. The impacts are only now starting to emerge into full view—and the forming picture portrays an existential challenge, to be urgently overcome using all the creative ingenuity we humans have relied upon throughout our history to solve our problems.

Knowledge about the toxicity of man-made industrial chemicals is not new: we have understood it since the ancient Romans found that work in the lead mines was a death sentence; since eighteenth-century chimney sweeps developed scrotum cancers; since phosphorus rotted the jaws of matchmakers and aniline dyes poisoned chemical workers in the nineteenth century; since phosgene and mustard gas choked off so many young lives in World War I. Since the Great Smog of London, Agent Orange, Minamata, Seveso, *Silent Spring*, the Love Canal, Bhopal, *Erin Brockovich*, the Asian Brown Cloud and the Great Pacific Garbage Pool.

Nevertheless, something vast has changed.

Today man-made chemicals and their by-products are everywhere, in all that we do: they are to be found in homes, offices and factories, on farms, in clothing and bedding, in electronics, in cars, aircraft and ships, in construction and industry, in pest control, and in products that we put onto or into our bodies, such as cosmetics, medicines, food, drink, tobacco and drugs. Unlike our great-grandparents and all generations before them, we are now immersed in these synthetic substances 24/7, no matter where we live: the chemical by-products of heavy industrialisation have spread around the planet and their fingerprints are to be found from the remotest poles to the deepest oceans, from our living blood, to our grave, to the genes of our grandchildren.

In modern society the world over, synthetic chemicals are integral to our daily lives. There is no industry or activity of advanced civilisation where they are not used in some form or other, with the aim of improving our quality of life. They solve problems, protect, adorn, kill pests, save lives, improve efficiency and enhance convenience. An advanced society without such chemicals is almost unthinkable. They are a part of who we are—but in far more ways than most of us would suspect.

The US Environment Protection Agency (US EPA) lists more than 84,000 different chemical substances manufactured or used in the United States alone.[4] The US Agency for Toxic Substances and Disease Registry (ATSDR) says that 'more than 100,000 chemicals are used by Americans'.[5] The European Community estimates more than 70,000 manufactured chemicals are currently marketed in its member countries.[6] In Australia, the National Inventory of Chemical Substances lists 38,000 chemicals it says are used nationally.[7] The UN Environment Programme (UNEP) states: 'The exact number of chemicals on the global market is not known but under the pre-registration requirement of the European Union's chemicals regulation, REACH, 143,835 chemical substances have been pre-registered. This is a reasonable guide to the approximate number of chemicals in commerce globally.'[8]

However, these formal registers do not include many tens of thousands of substances unintentionally produced during mineral extraction and processing, energy generation and the use of engines, during land degradation and as by-products of manufacturing or waste disposal. These substances have been entering our living environment largely unmonitored and in many cases are as yet undetected. Purposefully manufactured chemicals are thus only the very tip of a much larger iceberg of contamination unintentionally generated by productive activity as we make and refine all the material things we need and value.

However, the sheer number of legally manufactured substances affords a glimpse of the scale of today's chemical production enterprise. The UNEP expects this output to grow three-fold by 2050.[9] The value of chemical production worldwide rose from $171 billion in 1970 to $4.1 trillion in 2010; it is projected to reach $6.4 trillion by 2020 and triple again in value by 2050. The US Agency for Toxic Substances and Disease Registry estimates that 'about 1,000 new chemicals are introduced each year'.[10]

By 2020, says UNEP, more than half of global chemical output will be concentrated in developing and newly industrialising countries, where it will be effectively beyond the reach of the law. In India and China, for example, chemical output is forecast to grow by almost two-thirds.[11]

Chemical flood

The UNEP estimates that of the 5.7 million tonnes of chemicals released in North America (United States, Canada and Mexico) alone in the year 2006, 1.8 million tonnes were substances rated as persistent, bioaccumulative and toxic to humans and animals.[12] A million tonnes of these chemicals are linked with, or have suspected links with, cancer; a further 857,000 tonnes are linked to, or suspected of, causing birth defects and genetic diseases.

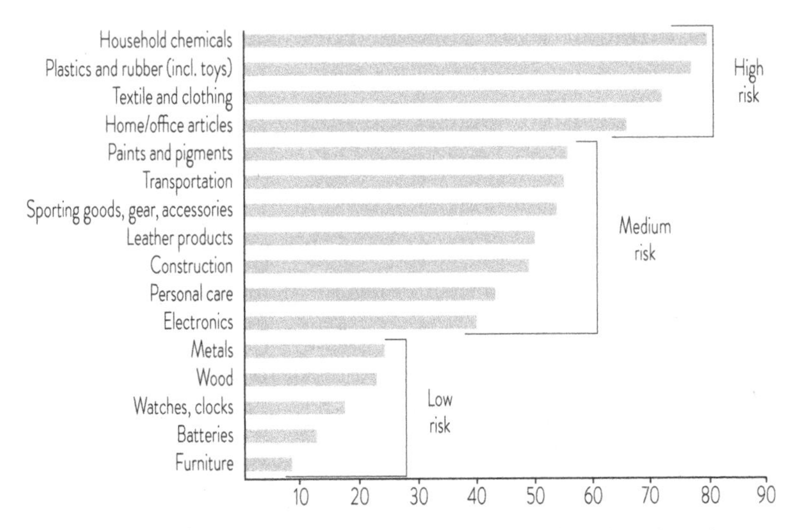

FIGURE 1 **Risks associated with common chemical-based products in our daily lives**

Source: United Nations Environment Programme (UNEP), 2012

Since North America produced around a quarter of the world's manufactured chemical output at the time, it may be inferred that the total release globally was between twenty and twenty-two million tonnes in 2006 and by the mid-2010s was of the order of thirty million tonnes, driven chiefly by rapid industrial expansion in Asia.

Another way to see the issue is that humanity is presently being carpet-bombed with around ten million tonnes of man-made toxins, potential carcinogens and hormone disruptors every year, and much of this release is cumulative, year on year. For each person on Earth, 1.4 kilos of things that can harm your genes, intelligence or health are now released annually. To give an idea of the scale of this release, it can be compared to the total chemical exposure suffered by the rural Vietnamese population during the 'Agent Orange' chemical defoliation campaign in the Vietnam War when an estimated 2.5 kilos of herbicides were released for each member of the exposed rural population per year. This activity has since been linked by doctors to 400,000 cases of death or maiming, and 500,000 cases of birth defects in Vietnam alone.[13]

These numbers are presented purely to give a sense of the scale of the chemical exposure of modern society and the individual while noting that, although many substances are considered harmless in small doses, it takes only a tiny quantity of a carcinogen to unleash a cancer. On present trends, global chemical output will reach 3.2 kilos of toxins per person per year by 2050.[14] In heavily chemicalised societies such as North America, the combined toxin output is already around 3.5 kilos per person per year.

Nutrient cascade

In addition to the thirty million tonnes of manufactured chemicals produced and released each year, humanity also concentrates and releases vast amounts of nutrients through worldwide agriculture, transport and manufacturing. Johan Rockstrom of the Stockholm Resilience Centre and colleagues[15] estimated that humanity injects 121 million tonnes of elemental nitrogen from the atmosphere annually into soil and water processes. We also discharge an estimated 8.5 to 9.5 million tonnes of elemental phosphorus into the oceans. Together these act as serious contaminants and greatly exceed the volumes naturally recirculating in the Earth system. In the case of nitrogen, they have already exceeded a dangerous boundary, a level which, in the opinion of the scientists, humanity ought not to transgress for its own safety. While not linked to large-scale human mortality, the release of these nutrients is slowly killing many of the world's rivers, lakes and an expanding area of ocean. This has so far led to the formation of hundreds of 'dead zones'—areas of water that have been stripped of oxygen due to nutrient pollution—and the resulting process of stagnation, leading to heavy losses of fish and other local water life.[16]

The release of nutrients is also closely tied to humanity's largest impact on the planet: soil erosion. With satellites revealing that half the Earth's land surface is already degraded, it has been estimated that around seventy-five billion tonnes of soil are displaced by farming, grazing, construction and development every year[17]—together with all the chemicals that bind to soil particles. While soil is not normally regarded as a contaminant, dust has a significant effect on human health, especially that of infants. Some dusts carry heavy metals and other toxins. It is also a reminder that we are mining the planet in order to feed ourselves—and that sooner or later, all mines become exhausted.

Mineral release

However huge it may seem, the output of the global chemical industry is dwarfed by that of the mining, mineral processing and energy sectors. To extract a single tonne of metal usually involves the crushing and refining of many tonnes of ore, creating a very large amount of waste material which is often toxic, because in the process of concentrating and extracting the desired metal or mineral other undesirable things also become concentrated and are released into the environment. These substances are customarily deposited in tailings dams or dumps, into rivers and oceans, whence they usually disseminate through the wider environment over a period of time. In total, the world's mines are estimated to produce more than six billion tonnes of ores a year,[18] most of which is waste and must be disposed of, either on land or at sea.

To take a single example, in producing the forty-five million tonnes of aluminium we need for cars, planes, cans, building materials and other products each year, the minerals industry generates around 120 million tonnes of an unwanted by-product known as 'red mud'. Though mainly consisting of sand, this substance is highly alkaline due to the Bayer process used to extract alumina from bauxite ore. Red mud is toxic to most plants and animals. Since aluminium production began more than a century ago, it is estimated the world has accumulated around three billion tonnes of red mud, for most of which no safe use or long-term disposal system has yet been devised.[19] Much of this substance is stored in dams, which are subject to unpredictable failure, while large volumes are routinely being dumped in the oceans. Aluminium is an invaluable product, but its story helps to illuminate the wider issue of the growing garbage pile of toxic by-products into which humanity is now sinking—most of which we simply (and conveniently) cast out of our minds.

In a second example, a group called Earthworks Action searched the records of the world's top ten mining companies and established that, together, they dispose of 180 million tonnes of tailings into nearby rivers and lakes, mainly in developing countries.[20] Since large mining firms are usually good citizens and take care of their reputations in order to protect their public sanction to mine, we may safely assume that the waste dumping practices of many thousands of smaller mining houses are in general, far more careless. Even the estimates of such massive quantities, it must be noted, do not include substances unleashed by the mining process itself, in which four or five tonnes of rock and soil must often be moved for every tonne of ore obtained. This material is then exposed to weathering and leaching, leading to the release of acids and other toxic substances.

Energy exposure

The largest mined product of all is coal, with around 7.7 billion tonnes produced every year to light and heat our homes, power industry, make steel and cement and supply many valuable and useful industrial products. However coal also leaves behind an extensive legacy of toxic contamination from its mining, processing, stockpiling, combustion, carbon emissions, fallout and fly-ash disposal.

Contaminants commonly found in coal include mercury, cadmium, radioactive elements, sulphur and nitrogen compounds, volatile organic carcinogens and other toxins. The carbon released when coal is burnt is the major driver of global climate change, according to a world scientific consensus. The mining and burning of coal also cause acid runoff and acid rain, which damage rivers and forests and are largely responsible for acidifying the world's oceans. By volume, the mining, burning and processing of coal constitutes the largest source of toxicity from any human activity and has been estimated to directly cause around 170,000 deaths a year.[21] Women are nowadays often advised by their doctors to avoid eating fish such as tuna during pregnancy to avoid poisoning their foetus with the mercury which these fish, due to their position at the top of the ocean food chain, concentrate at high levels in their flesh; grains and crops such as rice and green vegetables are also being contaminated. Most of this mercury, which has spread around the globe in water and air, is liberated when coal is burnt to power our cities and industries and circulates in the atmosphere, water and living creatures. But even today, few parents grasp the connection between exposing their baby to a proven nerve poison . . . and the simple, apparently harmless, act of turning on an electric light.[22]

In all, power stations and industry are estimated to add around 8100 tonnes of mercury into the world's air each year by burning coal.[23] Contrary to what many assume or hope, the mercury does not go away or become safe after a time: it accumulates year after year in living organisms and the environment. Even countries with strict clean-air laws cannot prevent this silent assault on the physical health of their own national populations from the atmosphere and food chain; the pollution continues to grow in countries where coal-burning is less stringently regulated.

Other fossil sources of power generation energy, including petroleum and gas, are estimated by the WHO to directly cost a further 130,000 human lives a year worldwide. To give an impression of the rate at which fossil fuel consumption continues to increase—thus the rate that accompanying toxins are liberated into our living environment—humanity consumed sixteen terawatts of energy in 2009, and we are on track to reach twenty-three terawatts by 2030. Of this energy, 87 per cent was derived from coal, oil and gas; only 8 per cent came from non-polluting, renewable sources.[24] On the face of it, energy demand driven by global economic growth means that the toll of death and disease caused by fossil fuels could potentially triple by the middle of the century.

In addition, the burning of fossil fuels liberates thirty-six billion tonnes of carbon into the atmosphere and oceans each year,[25] with potentially serious consequences for climate stability, temperature, sea levels, the ferocity and frequency of extreme weather events, the world food supply, human health and life expectancy. Carbon output, too, is on track to triple by mid-century and is unquestionably a major chemical signature imposed by the human species, which affects all life and the entire Earth system in a manner and to an extent that continues to be widely studied and reported in thousands of peer-reviewed scientific research papers each year.

Weaponised chemicals

Existing stockpiles of chemical weapons and huge amounts of nerve poisons dumped in the world's oceans over the last ninety years pose a continuing risk to human and environmental health. As of 2014, around 80 per cent of the world's declared stocks of chemical agents and almost half of its chemical weapons had been destroyed under the Chemical Weapons Convention (CWC)—but this still left 14,000 tonnes of nerve agents and ultra-toxins and 3.7 million loaded munitions,[26] or three-and-a-half times more than is necessary to kill the entire human race. However, only seven countries have signed the Convention and declared their stocks: significant additional stocks may be secretly held by seven other countries which have either not signed or ratified the Convention.[27] Concern persists over both the vast quantities of chemical weapons dumped in the world's oceans from 1918 to 1972, which have never been made safe and may leak out at some time in the future, and the safety of high-temperature incineration methods used to destroy existing stocks.

Nuclear legacy

The world uses around 70,000 tonnes of uranium a year in its 435 (soon to be 500) operational nuclear reactors while 5.3 million tonnes remain in reserves deemed economically viable to mine for use in existing technology—eighty years' supply.[28] The total amount produced globally since mining began a century ago is estimated at around 2.5 million tonnes.[29] The world's total nuclear waste inventory was estimated by the International Atomic Energy Agency in 2007 at 19.2 million cubic metres with a radioactive potential of sixteen million million million bequerels, or units of radiation. This comprises the waste from power generation, mining, nuclear weapons, decommissioned nuclear plants and nuclear medicine. Some of this waste remains dangerous for eons: iodine 129 for instance, has a half-life of 15.7 million years. Nuclear radiation is known to cause severe damage to genes and cells, giving rise to cancers, birth defects and life-threatening illnesses. The issue of the lack of long-term safe storage of nuclear waste remains largely unresolved, although many so-called solutions have been proposed.

Risky wastes

According to UNEP the world also produces some 400 million tonnes of other hazardous wastes annually which humanity struggles to dispose of safely.[30] Much of this is simply thrown into landfills, from where it leaches into the environment after a time, or else is dumped into developing countries where primitive recycling industries ensure the toxins end up recirculating in air, water, food, manufactured products—and people. In 1992 the Basel Convention was signed under which signatory nations agreed to restrict the international transport and dumping of toxic waste. Even today, this is thought to apply to less than 4 per cent of the hazardous waste generated worldwide each year.

Every day we throw things away and entirely forget about them: but this does not mean these things or their chemical components are gone for good. Many return to haunt us. Chemicals can leach out of landfill pits in groundwater or seep out as vapours. Some of them hitch rides on dust particles, leapfrogging around the planet in cycles of absorption and re-release in a phenomenon known as the 'grasshopper effect'. Many break down into relatively harmless substances, but others form more toxic compounds, and some may combine and recombine with one another over long periods, giving rise to generations of unintended by-products. As cities expand, buildings rise over old dumps, their inhabitants unmindful of the fact they are living close to potentially toxic zones. In many of the world's cities today, drinking water contains the lethal residue of yesterday's waste. While it is possible to clean up one or two toxic substances from a particular site, remediating the witches' brew of toxics produced in the stream of human chemical waste to a safe level is, generally speaking, beyond current technology or affordability, especially once these substances have escaped into groundwater and spread in the wider environment.

Finally, the world's water supplies—and through them the entire global food chain—are becoming increasingly contaminated by a wide range of chemicals capable of causing damage to the human brain, reproductive and hormonal systems. Substances capable of inflicting such damage are known as endocrine disrupting chemicals (EDCs), and their main sources include prescribed medical drugs and contraceptives, illegal drugs, pesticides and industrial petrochemicals found in both food and household goods. So far, more than 1500 of these chemicals have come to the attention of science; a few are now banned in Europe, but rarely elsewhere. To give an impression of the scale of human contamination of the Earth system, Table 1 shows the main categories of man-made chemical emissions.

TABLE 1 Estimated global emissions of significant chemical substances due to human activity, per year

Substance	Emissions
Uranium	70,000 tonnes
Pesticides	2.4 million tonnes
Toxic and carcinogenic chemicals	10 million tonnes
Other industrial chemicals	20 million tonnes
Phosphorus	10 million tonnes
E-waste	50 million tonnes
Nitrogen	121 million tonnes
Hazardous waste	400 million tonnes
Metals	1500 million tonnes
Mineral waste	5000 million tonnes
Petroleum	4400 million tonnes
Coal	8000 million tonnes
Carbon (all sources)	36,000 million tonnes[31]
Soil	75,000 million tonnes

These examples illustrate that our contaminating emissions are far larger, more diverse and pervasive than most people imagine, and involve a very much wider range of activities than chemical manufacture or the use of pesticides. Virtually everything we do causes the release of chemicals into our living environment. Thanks to our insatiable demand for the goods of industrial society, this output is on track to double—and maybe triple—within our lifetime. And while the environmental protection agencies of developed countries demonstrate sound knowledge and effective policing of individual industries and the dangerous chemicals they produce or use, these agencies typically have a limited view of the combined chemical impact on their population from all sources—especially from beyond their borders—and few ways to monitor or prevent it. While many individual companies may do their best to curb their own toxic output, they do not as a rule display much awareness of their contribution to the combined worldwide burden of toxic exposure or disease. And even distant countries can affect our health, our lives and those of our children, no matter where we live, since the processes of dumping and leaking toxic substances affect the entire Earth system, as will be shown in the next chapter.

For the first time in the Earth's history, a single species—ourselves—is poisoning an entire planet.

CHAPTER 2
POISONING A PLANET

All living things—from one-celled microbes to blue whales—depend on Earth's supply of air and water. When these resources are polluted, all forms of life are threatened. Pollution is a global problem.

National Geographic Education

Michael Vecchione first went to sea as a cabin boy on a three-masted schooner in Maine at the age of sixteen. It was the start of a lifelong love of the oceans, their mysterious deeps and all that lives in them which led to his becoming one of the world's leading experts on cephalopods—squid, octopuses, cuttlefishes and nautiluses—and Director of the US National Marine Fisheries Service National Systematics Laboratory. Cephalopods, he explains, are essential not only to fisheries management but serve as a general indicator of the health of the oceans and to the future of the twenty eight species of toothed whales, narwhals, dolphins and other sea creatures which feed on them.

In 2003, Vecchione and colleagues from the US National

Oceans and Atmosphere Administration (NOAA) and Virginia Institute of Marine Science, were ploughing through the heaving seas of the western North Atlantic, running a large mid-water trawl between 1000 and 3000 metres (3300–9900 feet) down. In their net they brought up nine different species including short-finned squid, cockatoo squid, jewel squid, vampire squid and the large jelly-like octopus *Haliphron atlanticus.* Keenly aware of the accumulating pile of research showing that persistent man-made chemicals were being found in whales and dolphins around the world, the researchers decided to analyse their catch to see if it could help explain the mystery of where these substances were coming from.

The results were a shock. These creatures which had spent their lives in the lightless depths, far from human populations, were contaminated with flame retardants from synthetic furniture and fabrics; electrical transformer chemicals called PCBs (polychlorinated biphenyls) that had been banned for thirty years; traces of the pesticide DDT—also banned—as well as highly toxic antifouling chemicals known to cause hormone disruption (severely restricted) and various petrochemicals.

'The cephalopod species we analyzed span a wide range of sizes and represent an important component of the oceanic food web,' Vecchione said. 'The fact that we detected a variety of pollutants in specimens collected from more than 3000 feet deep is evidence that human-produced chemicals are reaching remote areas of the open ocean, accumulating in prey species, and therefore available to higher levels of marine life. Contamination of the deep-sea food web is happening, and it is a real concern.'[1]

If there is one thing a deep-sea squid really doesn't need, its flame-retardants. Nor do lovers of seafood.

Local or global?

Pollution has long been viewed by society, governments and much of the scientific establishment as a local issue. Worldwide, it is estimated there are more than five million potentially contaminated sites,[2] with Europe having at least 1.7 million of these.[3] Yet the very term 'contaminated site' seems to imply there is a distinct boundary to the danger zone: that the hazards can somehow be walled off from society by putting an actual fence around them. However, new research is rapidly accumulating to indicate that the term 'contaminated site' is misleading: that no site is an island, completely isolated from the rest of its surroundings, and that most contaminants are, to some extent, perpetually in motion.

Man-made chemicals have been shown by science to move constantly in both space and time. They travel on the wind, in water, attached to soil, in dust, in plastic particles, in wildlife, in traded goods and in—and on—people. They combine and recombine with one another and with naturally occurring substances to form new compounds—some more toxic, others less so; some more mobile and others locked up, docked or hemmed in for a time. While some compounds quickly disintegrate into innocuous components such as carbon and oxygen, others seem virtually indestructible as they continually recycle through water, air and soil—and especially in the global food web and the human food chain.[4] This process may persist for many decades in the case of metals and especially tough chemicals such as DDT. We are thus increasingly exposed not only to the toxic residues caused by our own consumption of material goods, but also those of past socieities. And we, in turn, heedlessly inflict our own toxic outpouring on our children and grandchildren—not only directly, but also indirectly, through damaged genes, as we will see.

Cancer from the sky

In the early 1970s, scientists realised that eighty-nine man-made substances, chiefly chlorine- and bromine-based compounds mainly used to cool refrigerators and air conditioners and as spray-can propellants, were destroying the ozone layer in the upper atmosphere—the gaseous layer which shields life on Earth against the bombardment of deadly ultraviolet-B (UVB) radiation in sunlight. UVB is a primary cause of skin cancers, eye diseases and wrinkly aged skin.[5] Research on these ozone depleters, collectively known as CFCs, provided us with the first irrefutable evidence that even comparatively small and thinly dispersed man-made substances can affect the entire planet, posing a risk to everyone as well as to many other life forms. Thus, a hairspray used in America or France may contribute to a melanoma victim's painful death in Australia by accelerating the destruction of the protective ozone layer over Antarctica that once shielded people in the southern hemisphere from UVB rays when they played or worked in the sunny outdoors.

Action to end the use of such ozone-destroyers began with the signing of the Montreal Protocol in 1987, but a quarter of a century later, the phase-out has not progressed as quickly as hoped, partly because of a thriving global black market in dirty old refrigerators and air conditioners, and partly because some CFCs last a very long time.

Disturbingly, the original CFCs have been largely replaced by other chemicals, which—despite their ozone-sparing qualities—have turned out to be highly potent greenhouse gases, contributing about 0.4 billion tonnes of additional CO^2-equivalent emissions to the atmosphere each year, with emissions growing at a rate of 8 per cent a year.[6] Humanity thus appears to be in the process of trading the risk of death from skin cancer for the risk of dangerous climate change that may threaten many aspects of our lives including, as it turns out, causing further ozone layer depletion and hence increased cancer risk![7]

This story of the ozone depleters is a clear hint about how difficult it may prove for us to escape the chemical treadmill, especially by substituting one compound for another, which may turn out to have equally injurious side-effects. Also, who in today's modern society is willing to forgo their refrigerator or air conditioner—even though their use is estimated by world health authorities to cost the lives of up to 65,000 people a year worldwide and to cause rapidly ageing skin in all who venture into the sun?[8]

Every breath you take

Air pollution has been a matter of civic concern since at least 1271 when England's King Edward I banned the burning of sea-coal in his capital, London, because of the smog it was causing. Indeed, the citizens of ancient Athens and Rome frequently grumbled about the *gravioris caeli* (heavy atmosphere) that hung over their cities from potteries, furnaces and domestic cooking fires.[9] What is relatively new, however, is the appreciation that air pollution is no longer confined to individual cities or industrial basins, but now flows freely around the globe—and that what is produced in India or China may be inhaled by individuals in Canada or Germany, and vice versa, with equally deleterious effects.

Global air pollution caused by unregulated industrial development especially in the emerging economies—as well as activities such as the burning of forests and fossil fuels—has multiplied as a function of the growth in the human population and our burgeoning demand for energy, goods and services. Air pollution has widely documented impacts on our health. Japanese researcher Hajime Akimoto recounts: 'When the first measurements of high concentrations of CO [carbon monoxide] over tropical Asia, Africa, and South America . . . were made available in 1981 [from instruments located] on the space shuttle *Columbia*, it became clear that air pollution was an international issue.' In his groundbreaking report he went on to explain that global air pollution by ozone was now jeopardising agriculture and natural vegetation worldwide and having a strong effect on climate. During the 1990s, nitrogen oxide emissions from Asia overtook those of North America and Europe, and will continue to exceed them for decades to come.[10]

Air pollution now moves freely around both the Earth's hemispheres and across its largest oceans and continents, blotting out our view of the sun, moon and stars from many regions and producing ugly phenomena such as the Indian-Asian Brown Cloud, a vast grey-brown stain that hangs in the skies over much of southern Asia and consists mainly of soot particles and other pollution from various sources such as wood stoves, forest fires, motor vehicles, factories and coal-fired power plants.[11] As *The Economist* (an international business publication) vividly describes:

> The fetid smog that settled on Beijing in January 2013 could join the ranks of . . . game-changing environmental disruptions. For several weeks the air was worse than in an airport smoking lounge. A swathe of warm air in the atmosphere settled over the Chinese capital like a duvet and trapped beneath it pollution from the region's 200 coal-fired power plants and 5m[illion] cars. The concentration of particles with a diameter of 2.5 microns or less, hit 900 parts per million—forty times the level the World Health Organization deems safe. You could smell, taste and choke on it. Public concern exploded. China's hyperactive microblogs logged 2.5m[illion] posts on 'smog' in January alone. The dean of a business school said thousands of Chinese and expatriate businessmen were packing their bags because of the pollution.[12]

Air pollution alone is slicing five years off the life expectancy of the average Chinese, according to the US National Academy of Sciences.

According to the World Health Organization (WHO), outdoor air pollution is directly implicated in about 1.3 million deaths per year—the majority being children—which are mostly avoidable, an annual death toll which is increasing at the horrifying rate of 16 per cent per year.[13] Air pollution is a disease of prosperity, suffered chiefly by citizens of those regions which are emerging most rapidly out of poverty and into middle-class status. However, it is slowly but surely encircling the planet—and each one of us is copping it.

An even more serious problem is indoor air pollution, which is estimated by WHO to claim about two million lives every year.[14] Indoor air pollution has two main sources: wood, coal or kerosene fires in the developing world; and toxic fumes emitted by substances found in most homes and cars in the well-off world: furniture, wall and floor coverings, building materials, paints, plastics, foam rubber, bedding, pesticides and other oil-based products. And yes, that lovely 'new car' smell is toxic and bad for your health.[15] Both forms of pollution—indoor fires in the developing world and toxic fumes in the developed world—are linked to cancers and crippling lung diseases. In a disturbing development, it has become clear that toxic vapours from chemicals used to treat furnishings to prevent fire (flame retardants)—some of which are known to cause cancer—are now being absorbed into the bodies of millions of mothers and passed to their unborn and newborn babies. Since modern urban citizens spend between 90 and 95 per cent of their time indoors, their exposure to polluted indoor air is higher and much more prolonged than to outdoor pollution. This is ironic because opinion polls consistently show much higher levels of public concern about *outdoor* air pollution—probably because it is visible, whereas indoor pollution is not. Sadly, many people still seek refuge from outdoor smog by going indoors—where the invisible 'smog' may in fact be more poisonous. Furthermore, perhaps for the same reason, there is in general less government action to prevent indoor pollution than outdoor, and to properly regulate the industries which manufacture risky products.

Polluted poles

The polar bears which roam the Arctic would appear to inhabit one of the cleanest and least-polluted places on Earth, far from the urban grime of industrial society. Not so. In 2012, a team of researchers from Denmark, Greenland and Canada reported they had found the same types of flame-retardant chemicals in adult Arctic bears and their offspring over a period of twenty-seven years, with adverse impacts to the bears' health.[16] The levels in the animals' bloodstreams had more than doubled during the quarter-century of monitoring—yet wild bears are not directly exposed to toxic vapours from modern indoor furnishings and synthetic fabrics, so how did they become contaminated? The answer most probably lies in the gradual, planet-wide escape and dispersal of these substances in air and water from more highly polluted urban environments, then their precipitation in rain and snow, or their washing down rivers and concentrating in fish and other animals in the bears' (and our own) food chain. Similarly, Italian researchers described the finding of significant amounts of industrial mercury, mainly sourced from coal-burning in Europe and Asia, amid the Arctic snows, calculating that some 270 tonnes of the poisonous metal showers over the Arctic yearly.[17]

At the opposite end of the globe, in Antarctica, the impact is even more striking as the main source of pollution lies in the industrial heartlands of the northern hemisphere—yet, counter-intuitively, the pollution has somehow crept as far south as is possible on planet Earth. As Australia's Antarctic Division reports: 'Minute traces of man-made chemicals used in other parts of the world are now being detected in the snow that falls over the region. Some of these chemicals can become concentrated in the bodies of local wildlife, such as seals, penguins and whales, and can be harmful to these animals in the long-term.'[18]

In light of the destruction of the Antarctic ozone layer by chemicals largely manufactured in the north, this news ought not to surprise us, yet it still comes as a shock to discover that what many regard as the last 'pristine' environment on Earth is already unavoidably besmirched. Pollution in the Antarctic came to global scientific attention in the late 1970s, following which a long series of studies gave rise to growing concern. In 2006, a meeting of Antarctic Treaty Nations called for a report on industrial pollution of the only industry-free continent on Earth. The International Council for Science commissioned researchers from the universities of Venice and Pisa to carry out a systematic review which brought togther the findings of the main scientific papers published on the subject. They reported in 2009 that consistent levels of persistent organic pollutants (POPs), including pesticides, had been found in the atmosphere, sea water, ice, snow, marine environment and wildlife. By backtracking, the scientists concluded these pollutants had originated from as far afield as northern Europe and Russia, and had probably mostly reached the southern polar regions through the air. Significant levels of PCBs and hydrocarbon pollution were found in the surface waters of the Ross Sea, which lies off Antarctica south of New Zealand. Industrial chemicals were also found in krill, plankton, fish, penguins, birds, seals, whales and killer whales. These were generally found at lower levels than in similar animals living closer to human populations, but were widespread in the Antarctic wildlife and environment nonetheless.[19]

To underline the point, the Stockholm Convention says that POPs:

> concentrate in living organisms through another process called bioaccumulation. Though not soluble in water, POPs are readily absorbed in fatty tissue, where concentrations can become magnified by up to 70,000 times the background levels. Fish, predatory birds, mammals, and humans are high up the food chain and so absorb the greatest concentrations. When they travel, the POPs travel with them. As a result of these two processes, POPs can be found in people and animals living in polar regions, thousands of kilometres from any major POPs source.[20]

On a somewhat different polar topic, British and American researchers studying polar ice cores reported in 2013 that the Antarctic ice cap is currently melting ten times faster than it did 600 years ago, driven primarily by a 1.6-degree-Celsius increase in polar temperatures—the result of carbon emissions from the burning of fossil fuels. The melting had reached dramatic levels in the past half century, they said.[21] Carbon is simply another pollutant released by human industrial activities in excessive quantities, and which is now having planet-wide effects. Antarctic melting is of particular concern because once this process achieves momentum, it will be hard—if not impossible—to reverse. If the entire southern icecap eventually melts it will raise global sea levels by some 65 metres, inundating the world's port cities up to the twentieth storey of tall buildings (though this may happen over the course of several centuries).[22] The eventual loss—however slowly—of many of the world's finest cities such as London,

New York, Shanghai, Kolkata (Calcutta), Singapore, Sydney, Tokyo, Venice and Amsterdam along with large parts of Rio de Janeiro and San Francisco is no trivial matter and illustrates how uncontrolled pollution by industrial societies will eventually return to impact all civilisation, even though its first-round effects are in places many thousands of kilometres distant.

The Earth's highest point, Mount Everest, is sometimes referred to as the 'third pole' by virtue of its altitude and remoteness. In 2006, American scientist Bill Yeo climbed the mountain's Rongbuk Glacier and up its Northeast Ridge as high as 7752 metres (about 1100 metres below the summit), collecting samples of soil and freshly fallen snow for analysis. The results were an unpleasant surprise—the snow held traces of arsenic that were above the safety limit set by the US Environment Protection Agency (US EPA) for drinking water, with the levels rising with the altitude. The samples were also found to contain risky levels of cadmium.[23] While the origin of these poisonous substances is not entirely clear, their levels were higher than in the local rock and three to four times higher in Everest snow than in fresh snow from Antarctica, causing the scientists to suspect Asian industrial air pollution and dust from degraded farming landscapes as the most likely local pollution sources.

The discovery was not only a shock to mountaineers—who drink Everest's melted snow—but also has serious implications for millions of people around the world. High, icy mountains such as the Himalayas, Hindu Kush, Rockies, Andes, Urals and other great chains represent one of the largest stores of fresh water on the planet, the so-called 'water tower': each year their snows and glaciers melt, sending vast volumes of water down the rivers onto the plains below where people use it for drinking, washing and to grow food. No law protects people anywhere against this insidious form of pollution.

The toxic emissions of heavily populated areas of the globe thus return to bedevil humanity a few years later in a never-ending cycle, augmenting the impact of chemicals that already exist in the air, water and food of our cities.

The inescapable message of both these examples is that what we do as consumers, knowingly or in ignorance, comes back eventually to harm us.

Distant domains

For several decades scientists have been recording disturbing and significant traces of industrial pollution in marine animals such as whales, dolphins, fish and sea birds, often sampled many thousands of kilometres from urban civilisation. For instance, in 1997 Heidi Auman and colleagues reported finding highly toxic PCBs, as well as traces of the pesticide DDT, in albatrosses—chicks, adult birds and even eggs—on the remote Midway Atoll in the middle of the Pacific Ocean. The levels they found were comparable with those measured in water birds on the American Great Lakes, which are surrounded by industry: levels close to the concentrations which cause chicks to die in the egg.[24] At the time it was thought this might be due to the dumping of industrial waste into the Pacific, and while this is likely it is now thought these highly toxic substances are also circulating in water, wind, rainfall and the food web, eventually reaching even the most remote islands in the world.

In a second study in 2002, Derek Muir of Quebec University and colleagues reported much higher levels of pesticide residues: up to sixty times higher in albatrosses from the north Pacific than from the south Pacific, the difference being due to the fact that the northern bird population fed off the American coast, close to centres of industry, whereas the southern fed around New Zealand and in the remote Southern Ocean. Nevertheless, all the birds were found to be polluted to some degree.[25]

A similar gradient in pollution was observed in the early 2000s in harbour seal pups, by US and Canadian scientists who sampled along the coasts of British Columbia and Washington State from remote Queen Charlotte Strait to moderately industrialised Georgia Strait, to heavily industrialised Puget Sound. The pups from Puget were 'heavily contaminated' with PCBs and other organic chemicals, while those from the more remote places were less so, causing the scientists to suggest that harbour seals might be a useful 'canary in the coal mine' for the effects of heavy industry on the seas.[26] The seals also demonstrate the self-evident fact that while remoteness is no protection, the closer you get to heavy industry, the more contaminated you are likely to be. And most humans live very close to the source.

Typical contaminants of marine fish found during scientific studies include lead, cadmium, mercury, dioxins, PCBs, polycyclic hydrocarbons, flame retardants, melamine, marine biotoxins, histamines and radionuclides:[27] many of these contaminant classes originate from human industrial activity. While a great many scientific studies reveal that the contamination levels are mostly below the level of concern, researchers have expressed particular caution over the load of heavy metals and organochlorine pesticides in long-lived top predator fish, such as swordfish, tuna, sharks and rays, especially for communities where there is high seafood consumption. Even where fish contaminant levels may be low, there is growing concern that many of these contaminants may act as endocrine disruptors, meaning that they can potentially affect human health and development—especially of the brain, central nervous system, endocrine (hormonal) system and reproductive systems—at minute levels which are far below the toxicity limits set by regulators.

Even the mud on the seafloor is becoming poisonous, with disturbing consequences for marine life—and for the people who eat it. Ocean conservation body SeaWeb comments: 'Toxic contaminants lead to a severe reduction in the diversity of bottom-dwelling organisms that live in affected estuaries or coastal regions. And adverse effects can spread, via the food chain, to fish, birds, and mammals that feed on contaminated sea life.'[28] The Norwegian Government explains:

> Pollutants in sediments can spread to the surroundings. They may spread from the sediment to water, re-suspend when sediments are disturbed, or be absorbed by benthic organisms (bioaccumulation). Because of these mechanisms contaminated sediments may continue to release hazardous chemicals to the surroundings long after the land-based sources of the pollution have been eliminated. As a consequence contaminated sediments can have serious effects on living organisms and ecosystems.[29]

In marine organisms, POPs and heavy metals may cause higher mortality, reduce growth or disturb reproductive processes.

Because many contaminants accumulate in food chains, they can also affect human health—the lesson of Minamata.

Even in the dwindling wilderness of the Amazonian rainforest, the claws of contamination are reaching out. Extensive pollution by methyl mercury from thousands of small gold workings in the region has been recorded by scientists, affecting both the fish in the rivers and the people who rely on them for food, posing a threat to their health—again, as at Minamata.[30]

These, and many thousands of similar scientific and official reports, paint a disturbing picture of wildlife both on land and at sea becoming increasingly and extensively contaminated with man-made substances, and lend support to a view that many otherwise inexplicable population crashes and disappearances—such as the worldwide declines in frogs and honey bees—may be attributable, at least in part, to the spread of man-made chemical contamination. Besides the effects of direct poisoning, our pollution is thought by wildlife scientists to be having a more subtle impact on affected species by stressing their health and fitness to an extent that leaves them more susceptible to diseases, predators or environmental pressures—and by causing genetic damage, with long-term potential consequences for the survival of many species.

Out of sight, out of mind

In 1978, high rates of cancer and birth defects began appearing among residents living near the Love Canal in upstate New York. The problem was eventually tracked down: highly toxic industrial chemicals were leaching out of a nearby landfill and entering the local groundwater. State health commissioner David Axelrod presciently described the event as a 'national symbol of a failure to exercise a sense of concern for future generations'.[31] The failure persists: fifteen years after the celebrated revelations—made famous in the movie *Erin Brockovich*—that polluted groundwater in the California town of Hinkley contained chromium VI, the same cancer-causing contaminant was also found in the tap water of thirty-one out of thirty-five US cities tested in 2010, including Washington, Los Angeles, New York and Miami.[32]

In barely three decades, this failure has emerged in global proportions: today, many of the world's inhabitants now live on top of water so fouled by man-made substances that it is unsafe to drink. A British report states, 'In most countries the majority of groundwater under cities is contaminated beyond acceptable levels, and therefore cannot be used for drinking and other household purposes.'[33] In China, the groundwater underlying 90 per cent of the nation's cities has been described as 'polluted or severely polluted' by the China Geological Survey. Chinese government estimates noted 64 per cent of the nation's cities had severely contaminated water[34] and a staggering 100,000 Chinese were dying annually of chemical-related water poisoning. The Indian government has conceded that one-third of its districts are affected by groundwater that is undrinkable.[35] Another study estimated 140 million people in seventy countries are affected by arsenic contamination of drinking water; this occurs when groundwater pumping lowers the water table, drying out arsenic-rich sediments. The poisonous metal is then mobilised into a soluble and more highly poisonous form, arsenic III, which enters household wells and is used for drinking and cooking when seasonal rains soak into the ground and reflood the toxic layers.[36]

Groundwater is the one of the earth's largest natural resources, accounting for 97 per cent of the planet's total freshwater supply. It feeds most of our rivers and lakes, supplies 40 per cent of our water needs and grows nearly half our food. Groundwater can travel for tens or even hundreds of kilometres underground and is recharged by rainfall on scales ranging from days to millions of years. Consequently, if it becomes contaminated, the contamination can travel large distances in both space and time. The main sources of groundwater contamination are leaky landfills, hazardous waste disposal, illicit industrial discharges, seepage from old petrol stations, fuel dumps, factory sites and gasometers, mining and tailings dams, 'fracking', oil and chemical spills, fire-fighting chemicals, dry cleaners and mechanical workshops, badly managed sewage systems, medications, city runoff and farm chemicals. Because it is intimately connected to surface water, groundwater quality inevitably affects drinking supplies even when these are not drawn directly from wells. Once polluted, groundwater is exceptionally difficult and costly to clean up, unless the pollution is confined to a small area or a single source.

Stagnant seas

Something deeply disturbing is taking place in the world's oceans and estuaries: hundreds of dead zones are forming: areas devoid of oxygen and the sea life it supports. In recent decades the number of these aquatic black spots has risen steadily. As of 2013 there were 479 such sites reported, distributed along the most populous coastlines of Europe, Asia, the Americas and Australia.

Dead zones are not new. The first one appeared in the 1850s when industrialisation killed the Mersey River in the UK. But since that time they have metastasised, steadily and remorselessly invading all the oceans and seas most affected by human activity on land. Like the ominous blotches on a cancer patient's X-ray, you can follow their spread on the map in Figure 2.[37]

The cause of dead zones is well understood: their formation is driven by the avalanche of nutrients which humanity dumps in the oceans—from agriculture, sewage, leaky landfills, urban stormwater, soil erosion, industrial and vehicle emissions. This rich nutrient soup provides the food source for vast blooms of algae, and as these die off they sink to the sea floor and decompose, causing blooms of bacteria which strip the essential oxygen from the water column, often resulting in fish kills—their most visible impact. Dead zones are also hastened by global warming which stratifies the water, trapping the stagnant water and preventing it from mixing with the oxygen-rich surface layer.

The biggest contributors to these stagnant regions are 110 million tonnes of nitrogen, ten million tonnes of phosphorus and other nutrients which humans unleash upon the global ecosystem every year as we try to feed ourselves. If we seek to double our food production via traditional agriculture—as many say we must do in order to feed ourselves in the 2060s and beyond—the flood of nutrients will also double. That release, in turn, will spawn more and larger dead zones, like the zone which currently affects 22,000 square kilometres of sea at the mouth of America's Mississippi River and is driven by activities such as farming in the river catchment.

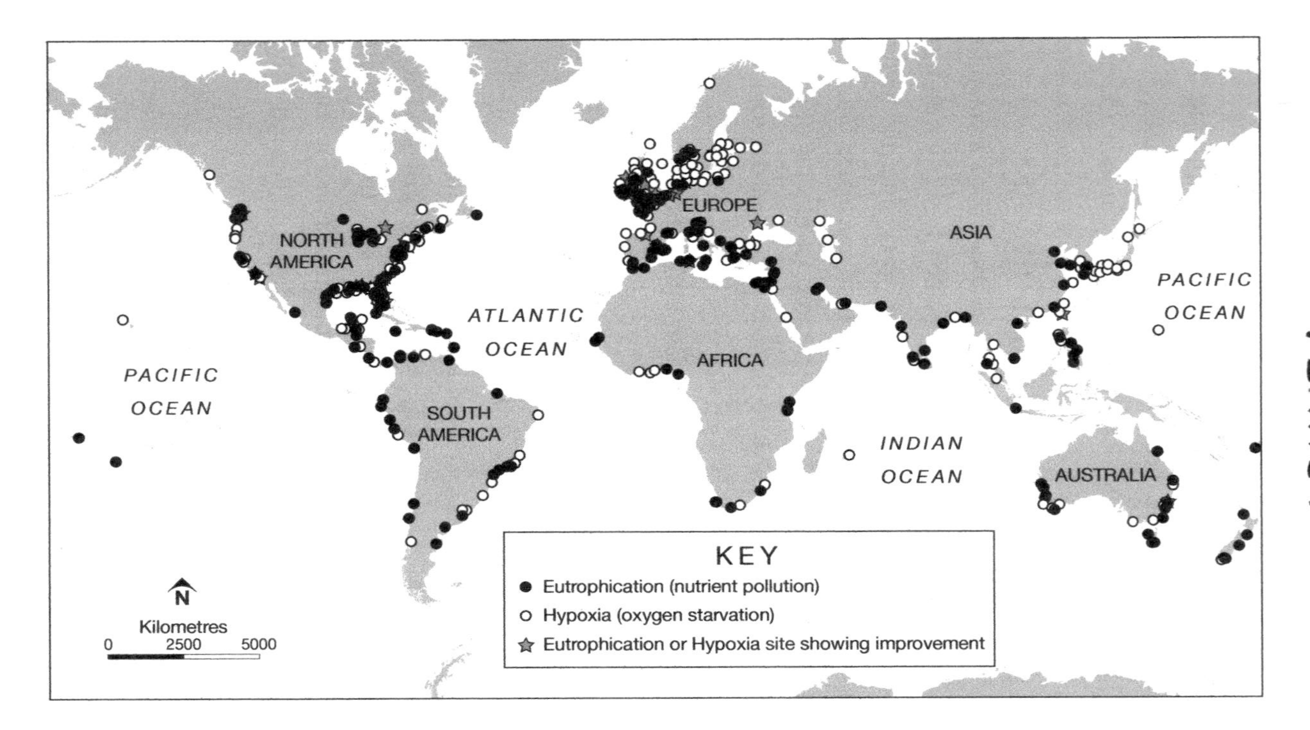

FIGURE 2 **World map showing areas affected by nutrient pollution**

Source: World Resource Institute (WRI), 2013

Dead zones are another unsettling manifestation of the human planet-wide chemical impact, even when we undertake something as innocent and well intentioned as farming. They are a plain warning that the 'leaky' production systems we have employed for centuries for food, mineral or industrial production cannot continue into the future and will have to change.

Another vivid example of planetary pollution is the 'Great Pacific Garbage Pool', an enormous gyre—or circulating current—of plastic trash, particles, oil and petrochemicals that swirls around the centre of the North Pacific, the accumulated result of decades of thoughtless disposal by Asia, the Americas and shipping. First reported in the mid-1990s, the patch has been variously estimated to cover an area from 700,000 square kilometres (270,000 square miles) to 15,000,000 square kilometres (5,800,000 square miles) and contain three million tonnes of plastic. This plastic, which has been shown to extensively contaminate and kill Pacific wildlife, especially seabirds, has a second, more ominous aspect. According to a study by Lorena Rios and colleagues from the University of the Pacific, California, the plastic fragments can also absorb contaminants such as PCB, DDT, PAHs and other POPs, and redistribute these many types of chemicals around the world's oceans.[38] Plastic waste is thus not only ugly and injurious to wildlife, but also serves as a vector for the continued recirculation of man-made toxins around the planet. Eventually much of this dangerous detritus washes up on coastlines or enters the marine—and hence human—food chain.

While the North Pacific Garbage Pool has commanded the lion's share of public attention, similar large garbage patches have been reported from the Indian Ocean and North Atlantic, and researchers consider it likely they are presently forming in all of the world's major ocean gyres.

Death of a giant

The greatest of all living organisms on Earth, and the only one that can clearly be seen from outer space, is Australia's Great Barrier Reef. Extending for more than 2600 kilometres, it is among the seven wonders of the natural world, a UNESCO World Heritage Area and an annual magnet for 1.8 million tourists from around the globe who come to marvel at the beauty and complexity of its corals, fish and other life.

Yet the Great Barrier Reef is already half gone—and in less than thirty years. 'This finding is based on the most comprehensive reef monitoring program in the world. The program started broad-scale surveillance of more than a hundred reefs in 1985 and from 1993 it has incorporated more detailed annual surveys of forty-seven reefs,' says Dr Peter Doherty, a Research Fellow at the Australian Institute of Marine Science. 'Our researchers have spent more than 2700 days at sea and we've invested in the order of $50 million in this monitoring program,' he says. 'The study shows the Reef has lost more than half its coral cover in twenty-seven years. If the trend continues coral cover could halve again by 2022.'[39] Coral experts such as the University of Queensland's Professor Ove Hoegh-Guldberg warn that without significant change in human impacts, the reef will be 95 per cent lost by 2050.[40] The reason is that man-made impacts are striking it faster and harder than the coral organisms can cope with, adapt to or recover from.

The AIMS study attributed the loss of coral cover to three primary causes: intense tropical cyclones, four population explosions by the coral-devouring Crown-of-Thorns starfish and severe coral bleaching events. Beneath these causes, however, are a host of drivers mostly related to human chemical emissions. Global warming, which is driven by fossil carbon emissions, raises sea water temperatures to intolerable levels for corals, causing massive bleaching events; it is increasingly implicated by science in more intense cyclonic (also known as typhoon or hurricane, depending on where you live) activity which causes direct destruction of shallow reefs. The worst impacts have been along the continent's coastal fringe where corals have been blanketed by eroded sediment released by farming, grazing, mining, dredging and development on the land. In this murky outflow are nutrients, fertilisers from farming, urban runoff and eroded soil, which cause seaweeds to blossom, smothering attempts by the corals to recover. The loss of reef fish, which normally keep the weeds in check, adds to this cycle.

Then there are the direct chemical impacts to the reef such as pesticide runoff from sugarcane farming, fruit and vegetable production and grazing, which destroy coral embryos even at very low levels of concentration.[41] Other adverse agents include substances like mercury from coal burning, oil spills and hydrocarbon runoff, pollution from towns and cities, antifouling agents used by shipping, and sewage. These in turn are linked by some scientists with an observed upsurge in viral and bacterial diseases of corals, which take hold when the corals are already stressed by factors such as toxins and rising water temperatures.

While Crown-of-Thorns starfish outbreaks occur on pristine Pacific atoll reefs, unaffected by human activity, they generally cause far less damage and the reefs soon recover. In areas affected by humans, however, they prove far more devastating. Scientists are still clarifying the exact mechanism but a favoured theory is that overfishing reduces the number of baby fish that would normally prey on starfish eggs and larvae, triggering a population boom. This is then sustained by the high level of runoff nutrients in the water which feeds the marine algae and plankton, providing food for the baby starfish, with the result that far more survive to adulthood and eat out the corals like a plague of undersea locusts.

A third of the world's fossil carbon emissions dissolve into the oceans, making them slowly but surely more acidic. This makes conditions more hostile for organisms like corals, which need calcium to build their skeletons. Evidence from prehistory suggests that acidification played a key role in five previous mass extinctions of corals seen in the fossil record, as Dr Charlie Veron explains in his book *A Reef in Time*.[42]

The fate of the Great Barrier Reef illustrates the potent synergy of a range of human chemical emissions to harm the world's largest organism—carbon (causing warming and acidification), sediment, nutrients, pesticides, mining and construction runoff, urban and industrial pollution—even when the actual levels of particular chemicals are very dilute. If together they can all-but kill off the biggest living thing on Earth, what effect are they having on the rest of life, and on ourselves?

The six pathways

The examples in this chapter illustrate that there are six main routes by which man-made pollution moves around the Earth:

1. Dissolved or as particles in water, including rivers, lakes, groundwater, rain, snowfall and ocean currents.
2. As airborne vapours, gases, microscopic chemical particles or attached to dust particles.
3. In the bodies of living animals and in plants.
4. Via the food chain, which has become contaminated by (3) and by the intentional use of pesticides, chemicals in packaging, food additives etc.
5. In manufactured goods—traded, transported and used by humans, deliberately or unintentionally, and in their disposal as waste.
6. In humans, being passed from mother to baby in the womb's blood and in breast milk, or from parent to child in damaged genes.

These pathways not only describe the planet-wide dispersal of man-made pollution, but also highlight the near-impossibility of controlling its spread anywhere except at its source—or very close to it. And while the Earth is a large place and pollution is still fairly dilute, pollution levels will likely grow several-fold by the middle of this century under the pressure of rising demand for food, housing, energy, industrial goods and services, as billions of people move out of poverty and towards a middle-class living standard, in which their consumption of polluting products is set to grow dramatically.

Tipping points

These many examples of the harms of pollution—and thousands more like them documented in the scientific literature—also demonstrate that man-made contamination is now a universal factor which impacts all nations and the health and wellbeing of every human being—no matter where they live—and indeed, most flora and fauna on Earth.

The UN Environment Programme says:

> Environmental effects of the chemical intensification of the national economies are . . . compounded by the trans-boundary movement of chemicals through the air or water. In some countries this occurs because they lie downriver or downwind from the polluting industries of neighbouring countries. In other countries, the runoff of pesticides and fertilizers from agricultural fields or the use of chemicals in mining in neighbouring countries, may leach into ground water, or run into estuaries shared across national boundaries. Throughout the globe, atmospheric air currents deliver chemical pollutants which originate from sources thousands of kilometres away.
>
> Whilst each chemical-intensification factor contributes to a small share of the environmental burden of each country and nation state, when combined, these together can form an increasingly significant and complex overall mix of chemicals not present fifty years ago. As this chemical intensity increases, the prospects for widespread and multifaceted exposures of humans and the environment to chemicals of high and unknown concern also arise.[43]

This is an issue for which currently no nation, industry, corporation or society accepts full responsibility, even for its own share of the mess. It is an issue to which most individuals are simply blind, even when it comes to protecting their own children. It is as if humanity has taken the profound collective decision that it is more important for us to consume than to protect our health or offspring—a decision which carries stark existential implications.

Although there is now ample scientific evidence of the injurious impact of individual chemicals on humans and other life, there is, as Johan Rockstrom of the Stockholm Resilience Centre warns, 'no aggregate, global-level analysis'. In short, we simply do not have clear information about how serious the position may be. As a result, also unknown are the thresholds at which large-scale tipping points, either in human or ecosystem health, may occur.[44] A tipping point is a term used in science to describe what happens when a system switches from one state to another—for example, when a river changes from clean and clear to foul, turbid and dead; or when formerly healthy farmland, forest or grassland turns into a desert through unwise use. Tipping points can often be reached quite quickly, without much warning and are generally difficult—frequently impossible—to reverse. An example of a major chemical tipping point is when the entire Earth's climate shifts from cool to warm due to carbon dioxide emissions. Not having a clear understanding of other chemical tipping points which we may trigger through the planet-wide dissemination of our contamination is a crucial blind spot in our ability to manage our future and that of our home planet.

This lack of understanding foreshadows developments which, for all our skills, science and cleverness, we cannot yet foresee. Yet it also challenges our creativity and inventiveness to find a viable solution, as soon as we can.

CHAPTER 3

ARE YOU A CONTAMINATED SITE?

It is ironic to think that man might determine his own future by something so seemingly trivial as the choice of an insect spray.

Rachel Carson, *Silent Spring*, 1962

In a run-down hospital in Bihar, India, five-year-old Rashmi Kumari was fighting for her life. She was the only child of her family to remain alive; twenty-two of her school friends had already perished. From the report by journalist Rajesh Roy in *The Wall Street Journal*:

> Inside an ambulance, an Indian man cradles his dead daughter, who had fallen ill after eating a school lunch. Initial investigations suggest that organophosphorus, commonly used in farm pesticides, may have been mixed into the rice, beans and potato curry served at an elementary school in Gandaman, a village in the impoverished state of Bihar, according to Amarkant Jha Amar, medical superintendent at the hospital.

> The students became sick and suffered from vomiting, fainting and foaming at the mouth. More than two dozen victims are still being treated. The state's education minister, P.K. Shahi, said the school's cook noticed a strange color and foul smell from the cooking oil when preparing lunch. Children also complained about the food, he said. He said the cook informed the school's headmistress. After tasting the food, the cook also became ill and was hospitalized.[1]

News of the Indian school children's poisoning rang out around the world. Adding to the horror was the fact that the poison was sourced to a Government-sponsored school lunch program intended to help poor families by providing better nutrition for their children. Someone, it seemed, had put the food in old pesticide containers.

'Just another regrettable accident in a chemically ill-regulated country,' was the reaction of many in the West. 'It couldn't apply to us.'

Or could it?

Whose school lunches are really free of all forms of industrial chemical, toxin or pesticide? Whose children are free of such things? Who has seriously investigated?

* * *

Society is accustomed to thinking of pollution in terms of contaminated sites: areas of land too heavily affected by industrial leftovers to be safely used. Yet mounting scientific evidence from all around the world is revealing that today, each individual is their own contaminated site. From the moment of conception to the moment of death, we all now accumulate a potentially debilitating and sometimes lifelong store of poisonous substances that originate in our consumption and living patterns. Many of these substances can now be identified by a range of common, though expensive, tests. This is not a matter for belief, denial or ideology: if you doubt it, get your blood, hair, saliva and urine thoroughly tested.

The world was awakened to the risks of profligate use of toxic chemicals by American biologist and writer Rachel Carson, when she published her celebrated book, *Silent Spring*, in 1962. Carson's focus was on the excessive use of long-lived pesticides such as DDT in the food chain and environment at the time—yet pesticide use has since increased more than thirty-fold to around 2.36 million tonnes globally today.[2] Once employed chiefly in developed-country agriculture, the use of pesticides has spread around the world, driven chiefly by supermarket and food company demand for high volumes of cheap, uniform, blemish-free produce—with the result that few, if any, consumers of the modern diet and industrial goods would find themselves uncontaminated if tested.[3]

Yet pesticides, despite the high public focus and media scrutiny they receive, are in fact only the minute tip of the chemical iceberg, most of which remains out of sight and out of mind.

So immense has the task now become of measuring—in either the environment or in humans—the true extent of contamination from the tens of thousands of legal chemicals and tens of thousands more toxic substances produced as unwanted and unintended by-products by industry, energy, construction and mining, that most national regulators and governments do not even attempt it. Instead they confine themselves mainly to limited investigations of a small number of substances of immediate concern in certain subsets of the population.

This means that, at best, they are examining only a handful of pieces from a ten-thousand-piece jigsaw puzzle, and revealing only dissociated fragments of the image it portrays.

Polluted people

One of the best studies is by the US Centers for Disease Control (CDC), which regularly monitors for around 212 out of the 84,000 chemicals made or used in the US, using a sample of 2500 people—a group it says is representative of the general population of 320 million.[4] It selects which chemicals to monitor on the basis of:

> scientific data that suggested exposure in the U.S. population; the seriousness of health effects known or suspected to result from some levels of exposure; the need to assess the effectiveness of public health actions to reduce exposure to a chemical; the availability of a biomonitoring analytical method with adequate accuracy, precision, sensitivity, specificity, and throughput; the availability of adequate blood or urine samples; and the incremental analytical cost to perform the biomonitoring analysis for the chemical.

Even such a large and carefully conceived survey plainly leaves plenty of gaps: if it were to be compared with the coastguard's task, for example, it would equate with patrolling just 50 kilometres (31 miles) out of the USA's total 20,000 kilometres (12,383 miles) of coastline and hoping that would provide general warning of any incursion. It leaves a lot of scope for 'chemical Pearl Harbors'.

The CDC found the fire retardant PBDE—a substance known to affect nerve development and hormone production in humans—in the blood of virtually every American it examined. It found BPA, a compound found in plastic drink bottles and food cartons and thought to cause reproductive disorders and heart disease, in the urine of more than 90 per cent of those surveyed. It found measurable levels of the toxin and carcinogen PFOA (perfluorooctanoic acid) from non-stick cookware in 98 per cent. It found the lethal neurotoxin acrylamide—from either cooked potato chips or tobacco—was 'extremely common'. It found the substance perchlorate, used in the making of rocket fuel and explosives, in every person tested. It found the fuel additive MTBE in 'a high percentage' of the population. It also found 5 per cent of the population had levels of the toxic metal cadmium in their blood close to the level of concern.

The CDC notes that the presence of a certain chemical in blood or urine need not mean the health of the individual is at risk. However it also cautions that the threshold levels for adverse health effects of many chemicals are simply not known. And it is silent about the possible health effects of mixtures of chemicals and their breakdown products.

Acknowledging the vastness of the problem, the World Health Organization comments:

> The number of existing chemicals and their compounds is very large, and for many of them the health risks are not known. Chemicals can be the result of anthropogenic sources [caused or produced

> by humans] or occur in nature. Hazardous chemicals can reach our body through different routes (e.g. food, air, water) and cause a variety of health effects. Due to the many ways in which chemicals are used and released, the many exposure routes involved, and the different mixtures of chemicals present, the public health relevance of chemicals can be extremely difficult to assess.[5]

Exposure to and accumulation of chemicals continues in most people throughout their lives. To take but one example, the WHO considers that '*all people* [author's emphasis] have background exposure and a certain level of dioxins in the body, leading to the so-called body burden'. While this may not cause serious illness in most people, nevertheless the risks associated with dioxin exposure include skin diseases, altered liver function and cancers. The effects are far more dangerous for the unborn child. Dioxins are found throughout the world in the environment and they accumulate in the food chain, mainly in the fatty tissue of animals. WHO continues:

> More than 90% of human exposure is through food, mainly meat and dairy products, fish and shellfish. Dioxins are highly toxic and can cause reproductive and developmental problems, damage the immune system, interfere with hormones and also cause cancer. Due to the omnipresence of dioxins, all people have background exposure, which is not expected to affect human health. However, due to the highly toxic potential of this class of compounds, efforts need to be undertaken to reduce current background exposure.[6]

Concerned at some of the gaps in our knowledge, the Environmental Working Group (EWG), a not-for-profit American environmental NGO, ran eleven separate studies of its own, with the aim of finding out how polluted Americans are. It established they were affected, literally, cradle to grave. 'These projects, employing leading biomonitoring labs around the world, have together identified up to 493 chemicals, pollutants and pesticides in people, from newborns to grandparents,' it said.[7]

BPA babies

In an even more striking report in 2009, the EWG revealed that Americans now enter the world already laden with industrial chemicals: 'A two-year study involving five independent research laboratories in the United States, Canada and the Netherlands has found up to 232 toxic chemicals in the umbilical cord blood of 10 babies from racial and ethnic minority groups. The findings constitute hard evidence that each child was exposed to a host of dangerous substances while still in its mother's womb.'[8]

Nine out of the ten baby samples contained bisphenol A (BPA), an industrial petrochemical produced by the millions-of-tonnes annually in the production of polycarbonate plastics and epoxy resins: 'BPA has been implicated in a lengthening list of serious chronic disorders, including cancer, cognitive and behavioral impairments, endocrine system disruption, reproductive and cardiovascular system abnormalities, diabetes, asthma and obesity.' Major sources of BPA, and presumably the reason for its high prevalence, are its widespread occurrence in plastic drink bottles and food cartons, and in timber particle board and furniture made with epoxy glues. Bizarrely, most of the babies' blood also contained traces of chemicals used in explosives and rocket fuel, which EWG attributed to poorly maintained defence facilities allowing such chemicals to leak into groundwater, which supplies most of America's drinking water. Additionally, industrial chemicals—used to make cosmetics, detergents, soap and other scented household items—were found in seven out of ten samples.

'Each time we look for the latest chemical of concern in infant cord blood, we find it,' said Dr Anila Jacob, EWG senior scientist and co-author of the report. 'This time we discovered BPA, among other dangerous substances, in almost every infant's cord blood we tested.'

Ten babies in the EWG study may not sound many, but these findings sufficiently alarmed scientists the world over to spark a series of massive taxpayer-funded studies, several of which involve more than 100,000 infants, in countries such as Denmark, Norway, the United States, Britain, Germany, Japan and in the city of Shanghai in China. The study involves testing samples from the blood, hair, urine and saliva of babies and of their parents—as well as samples collected from their food, homes and local environment—for literally hundreds of different toxic substances then correlating the findings with health records. These are 'prospective studies', meaning that health researchers hope that by monitoring the child's chemical contamination levels throughout his or her early life, any health problems which show up later in life can potentially be linked back to previous recorded exposure to toxic substances. Researchers hope that together, these studies will reveal the first truly worldwide picture of the extent and sources of the body burden of pollution carried by our species.[9] However, since these studies cover the whole life of individuals from birth to thirteen (or even twenty-one) years of age, it will take at least another decade before clear study results emerge. And by that time there will be around 10,000 new man-made chemicals in the world for us to worry about.

Today's newborn infant has barely drawn its first few breaths before the chemical exposure continues, as soon as it begins to feed. A team at University of California Berkeley found pesticides and PCBs in *all* samples of mothers' breast milk it collected in the San Francisco Bay area and in the Salinas Valley.[10] The most disturbing feature of this study was that most of the pesticides found were modern, supposedly 'non-persistent' pesticide types—which were introduced specifically to replace banned 'persistent' organochlorine chemicals—and whose health effects remain 'largely unknown' according to the CDC. A Chinese study in the Guangzhou region reported similar findings to the UC Berkeley study: banned DDT, dioxins, organochlorines and other POPs were detected in mother's milk.[11] And in Europe, researchers reported dioxins, PCBs and other POPs in the breast milk of nursing mothers from fifteen different countries in four separate surveys between 1987 and 2007.[12] Encouragingly, the Europeans noted a decline over time in the presence of some banned substances, such as dioxins and DDT, in the milk in some countries—but, discouragingly, they reported a rise in the presence of new chemicals of concern, such as flame retardants. Together these various studies raise a strong possibility that mother's milk, that uniquely precious and revered fluid, is being contaminated on a worldwide scale—and that the banning of a few selected chemicals is not relieving the overall problem.

A major source of the contamination of newborns is 'air toxics'—volatile organic chemicals (known as VOCs) many of which are emitted by the home and its furnishings, then inhaled by the mother and passed to her baby via her bloodstream or milk. The US EPA broadly classifies air toxics as dangerous substances found in the outdoor air of big cities, however this view tends to downplay the far more insidious and serious issue of exposure to equally harmful things in homes and workplaces where we often assume we are safe from polluted air. The Australian Government explains: 'VOCs are emitted from some fabrics, carpets, fibreboard, plastic products, glues, and solvents, some spray packs and some printed material, paints, varnishes, wax, cleaning products, disinfectants, cosmetics, degreasing products, hobby products, fuels. Petrol stations are significant emitters of VOCs.'[13] Since modern urban citizens spend more than 90 per cent of their time indoors,[14] exposure to indoor air toxics is now regarded as a serious and lifelong potential source of health issues, as these toxics may be inhaled with almost every breath. Pregnant women and their unborn babies are especially at risk in the modern home.

Other potential sources of contamination of the foetus and newborn are the now-ubiquitous cosmetics and perfumed soaps. A separate study by EWG found an average of thirteen cosmetic chemicals in the bodies of twenty teenage girls—a group particularly vulnerable to hormone-affecting substances. Among them were phthalates, triclosan, parabens and synthetic musks, which have all been linked either to cancer or to hormonal disruption.[15] These same chemicals were also found in newborn babies, having reached them via the mother's bloodstream. Though the scientific data remain sparse, it nevertheless raises the spectre of serial contamination of successive generations of modern women in countries around the world where these products are now commonly used. Vanity, it seems, is toxic.

Cosmetic contamination

It might almost appear that the first thing the modern citizen does upon waking each morning is to try to cultivate a cancer. Most people shower or bathe themselves in chemicals from head to foot, many of which are suspected or known to be toxic, carcinogenic or allergenic. Today's shampoos, soaps, bath gels, cosmetics and lotions contain a cornucopian mixture of contaminants—many of them derived from petrochemicals—that add greatly to the total chemical exposure of both the user of such products and that of everyone around them, including babies and children.

The University of Toronto, which conducts one of the world's leading research programs into cosmetics, says:

> Exposure to perfumes and other scented products can trigger serious health reactions in individuals with asthma, allergies, migraines, or chemical sensitivities.
>
> Fragrances are found in a wide range of products. Common scented products include perfume, cologne, aftershave, deodorant, soap, shampoo, hairspray, bodyspray, makeup and powders. Examples of other products with added scents include air fresheners, fabric softeners, laundry detergents, cleaners, carpet deodorizers, facial tissues, and candles. We generally think that it is a personal choice to use fragrances; however, fragrance chemicals are by their very

> nature shared. The chemicals vaporize into the air and are easily inhaled by those around us. Today's scented products are made up of a complex mixture of chemicals which can contribute to indoor air quality problems and cause health problems.
>
> Susceptible individuals can experience a variety of symptoms, including headache, sore throat, runny nose, sinus congestion, wheezing, shortness of breath, dizziness, anxiety, anger, nausea, fatigue, mental confusion and an inability to concentrate. Although the mechanisms by which fragrance chemicals act to produce symptoms are not yet understood, the impact on all those affected can be quite severe, resulting in great difficulty in work and study activities.[16]

In view of all this there is growing pressure to limit, or even ban, the wearing of fragrances in some work places and public venues: perfume, it seems, could one day become about as politically correct as tobacco smoke or a seal-fur coat.

The EWG points out that the American government does not require health studies or pre-market testing of the chemicals in personal care products, even though just about everyone is exposed to them. It cautions that consumers tend to believe a great many myths about cosmetics.[17] 'The [US] Food and Drug Administration has no authority to require companies to test cosmetics products for safety. The agency does not review or approve the vast majority of products or ingredients before they go on the market . . . With the exception of color additives and a few prohibited substances, cosmetics companies may use any ingredient or raw material in their products without government review or approval,' the EWG explains. It publishes an extensive list of substances commonly found in cosmetics and body treatments which consumers are advised to avoid for health reasons when making their choices. These include things such as parabens, which are common in shampoos: 'Parabens are estrogen-mimicking preservatives, found in breast cancer tumors of 19 of 20 women studied. The CDC has detected parabens in virtually all Americans surveyed.' The EWG database, which contains 62,000 known cosmetics additives, lists eighteen different parabens and assigns to each a hazard rating.[18]

The US National Institute of Occupational Safety and Health found that 884 of the chemicals available for use in cosmetics had been reported to the US government as toxic substances—but their safety had not been officially assessed by the Food and Drug Administration, adds America's Cancer Prevention Coalition. 'In 1990, there were some 38,000 cosmetics-related [*sic*] injuries that required medical treatment in the U.S. That figure does not include the many people who use cosmetics and suffer from allergies, irritation, and photosensitization,' it adds.[19]

However, given the pleasure which these products widely evoke, it may be difficult for people who enjoy using perfume and cosmetics to appreciate they are in fact risking self-poisoning or exposing their neighbours in a manner similar to tobacco smokers. The toxicity that on average, people widely recognise in the cases of tobacco, alcohol, drugs and food is, in the case of the deceptively named 'personal care' products, blithely overlooked. Yet cosmetics use underlines a vital point about the nature of global contamination: this contamination arises not only from the unscrupulous actions of large corporations and the seemingly benign neglect of governments, but also from the combined wishes, preferences and demands of billions of consumers who are prepared to risk trading their health, and that of others—including their children—for the immediate gratification which such luxuries afford.

Breast cancer seems a high price to pay for great hair.

Food fears

In 2010, North Carolina mother Lisa Leake had what she freely admits was an epiphany. She was watching author and science writer Michael Pollan being interviewed about his latest book on *The Oprah Winfrey Show*. 'I was very intrigued and went on to read his book *In Defence of Food*. It was a huge wakeup call to me. I couldn't sleep at night when I was uncovering all this information . . . I was appalled because I thought I was feeding my family healthy food—but the foods I thought were healthy turned out to be highly processed.'[20] Like a good many innocent consumers, Lisa thought there were actual strawberries in strawberry syrup and lemons in lemonade, only to discover to her horror the flavourings were in fact made from chemicals. It was the start of an interest that became a fascination and then a crusade, in the form of Lisa's *Real Food* website and blog which has attracted hundreds of thousands of fans worldwide and provides advice about what 'real food' is and where to get it.[21] She converted her family's diet to one of whole, fresh foods, noting immediate improvements in health. Her infant daughter who had suffered constipation and asthma was freed of both ailments. 'We all had a decrease in sicknesses over the next year,' she said. Her own blood cholesterol reading plummeted spectacularly in the healthy direction. To show that it is indeed possible to eliminate many of the deleterious substances from our modern diet, she designed her 100-day Real Food Challenge, where families commit to eating mainly whole grains, fresh fruit and vegetables, locally raised meats and other wholesome foods that have not been processed or had substances added to them. She adopted Michael Pollan's advice: read the label first and never eat any food containing a substance you can't pronounce or you don't trust. She advises her fans to patronise the local farmer's market and a local bakery that uses wholegrain ingredients instead of the supermarket. What began as one concerned mother's obsession has given rise to a movement that is rapidly spreading around America and beyond.

Lisa is one among millions of consumers around the globe who are wising up—and shunning the modern industrial food supply, considering it a major source of lifelong personal contamination. Due to the extensive use of chemicals to grow, preserve, protect, process, sweeten, dye and flavour at least half of the world's food, the modern diet is increasingly recognised by scientists and health experts as risky. For example, the World Health Organization states:

> The contamination of food by chemical hazards is a worldwide public health concern and is a leading cause of trade problems internationally. Contamination may occur through environmental pollution of the air, water and soil, such as the case with toxic metals, PCBs and dioxins, or through the intentional use of various chemicals, such as pesticides, animal drugs and other agrichemicals. Food additives and contaminants resulting from food manufacturing and processing can also adversely affect health.[22]

The Union of Concerned Scientists adds: 'In pursuit of productivity, industrial agriculture degrades the air, water, and soil; damages fisheries and wildlife habitats; harms rural communities; poisons farmworkers; and undermines the natural resources on which future farmers depend.'[23]

To take but a single example, much industrially produced food today contains dyes and colourings, many of which have been scientifically linked to, or are suspected of causing, cancer and other health issues. Comments University of California LA molecular toxicologist Sarah Kobylewski: 'Dyes are complex organic chemicals that were originally derived from coal tar, but now from petroleum. Companies like using them because they are cheaper, more stable, and brighter than most natural colorings.' In her report *Food Dyes: A rainbow of risks*, she points out that Americans alone may be eating around 3000 tonnes of these dyes a year, most of which have not been properly health tested—except, in some cases, by the people who make them! Consumption of dye increased six-fold from 10 milligrams per person per day in the 1950s to 60 milligrams in the 2000s, making it likely that the consumption of carcinogens—which contaminate some common dyes—had also reached unsafe levels.[24] She pointed out that several US multinational food firms were selling the same food products in the US and Europe, with only the US version containing dyes. 'Most of those companies said that they don't use dyes in Europe because government has urged them not to—but that they would continue to use dyes in the United States until they were ordered not to or consumers demanded such foods,' she noted.

Among the most dangerous contaminants of the modern industrial food chain are cancer-causing dioxins, of which the EWG comments, 'the ongoing industrial release of dioxin has meant that the American food supply is widely contaminated. Products including meat, fish, milk, eggs and butter are most likely to be contaminated, but you can cut down on your exposure by eating fewer animal products.'[25]

Owing to the understandably high public profile of the food contamination issue, it receives extensive regulatory and media attention in developed countries, and much less in developing nations—although consumer awareness is growing quickly. However, even in the best-managed societies, this scrutiny tends to focus on one chemical at a time, rather than on the complex mixtures and synergies which inevitably occur within an individual's diet. We may thus be receiving an artificially reassuring picture, compared to the true extent of both public and personal health risks.

The complexity of the modern food chain makes it almost impossible for the average urban consumer to completely avoid pesticides and other chemicals, even if they mostly eat certified-organic fresh produce (which surveys have consistently shown is not always free of man-made chemicals). Some of these chemicals will sneak into the diet in chocolates, lollies (candy) and ices, soft drinks, tap water, bottled water, fast food, processed food, snack foods and so on, all of which are hard to obtain in pesticide-free forms. As a way of reducing the toxicity burden, the EWG monitors the contamination status of American fresh fruit and vegetables from year to year and advises consumers as to which ones are most contaminated and which are least. In its 2013 summary it reported:

> The most contaminated fruits are apples, strawberries, grapes, peaches and imported nectarines. The most contaminated vegetables are celery, spinach, sweet bell peppers, cucumbers, potatoes, cherry tomatoes and hot peppers. Every sample of imported nectarines tested positive for pesticides, followed by apples; 99 per cent of apple samples tested positive for at least one pesticide residue. The average potato had much higher total weight of pesticides than any other food crop. A single grape tested positive for 15 pesticides. The same was true for a single sweet bell pepper. Single samples of celery, cherry tomatoes and sweet bell peppers tested positive for 13 different pesticides apiece.

It found the cleanest produce to be pineapple, papaya, mango, kiwifruit, cantaloupe (rockmelon), grapefruit, corn, onion, avocado, frozen sweet peas, cabbage, asparagus, eggplant, sweet potatoes and mushrooms.[26]

Europe on the whole takes a much stricter view of the presence of pesticides and chemicals in food than does America, claiming to have outlawed some 750 substances that did not meet its criteria for causing no harm to humans or the environment. That still leaves around 250 chemicals for legal use on European farms.[27] A 2010 study which involved the analysis of 178 different pesticides in 500 different raw and processed foods, found that 2.8 per cent of all samples tested had residues of at least one chemical above those permitted under European law. The main offender was oats (5.3%), followed by lettuce (3.4%), strawberries (2.8%), peaches (1.8%), apples (1.3%), pears (1.3%), tomatoes (1.2%), leek (1.0%), cabbage (0.9%) and rye (0.2%). Seventy-nine samples were found to contain more than thirty different pesticides and to possibly present an acute risk to anyone who ate them. Among the samples of organic foods, 0.8 per cent exceeded the maximum permitted pesticide level. Overall, the authorities concluded, 'long-term exposure of consumers did not raise health concerns'—however, this opinion was based on food exposure alone, and did not include the exposure constantly accumulated by an individual via other physical routes such as drinks, medications, the skin or breathing.[28]

For the first time, the EU has also attempted a cumulative risk assessment, examining the exposure of consumers to combinations of different food chain chemicals; these findings are still at an early stage. And they do not include chemical exposure from other sources, such as the home, the highway or workplace.

The levels of chemicals in food in other parts of the world, especially developing and newly industrialising nations, are thought to be much higher because of several factors: lack of regulations, corruption and inadequate enforcement by the authorities, and the lack of experience of farmers in how to use pesticides safely. This is added to farmers' general attitude about pesticides: that when it comes to controlling pests, 'more is better'. There may also be extensive continuing illegal use of banned substances, such as DDT and other organochlorine pesticides, driven by cost pressures in the global food chain, as suggested by the Chinese survey of contaminated mother's milk. Investigating pesticides in imported produce, the EU survey reported that 'Products from South-East Asia still often violate limits.' It found that food imported from Thailand, Kenya, Uganda, China, India, Egypt and the Dominican Republic were the most frequently in excess of Europe's limits.

Owing to the globalisation of food and the unending quest by giant supermarket chains, fast food chains and food processing corporations for ever-cheaper sources of supply of uniform produce—no matter where it comes from or how it is produced—it is quite likely that the level of pesticides in the diet of ordinary consumers is rising, as local shops source ever more cheap food on their shelves from developing countries, and as more cheap, anonymous ingredients find their way into bulk-processed, manufactured, frozen and fast foods. Just because you may live in a well-regulated society does not make you safe from pesticide pollution from the developing world: such is the burgeoning river of today's globally traded food, and such is the universal movement and concentration of pesticides up the biological food chain and in air, water, plants, animals and people. Even people who hunt or grow their own food are not safe from chemical exposure: wildlife and urban soils are both widely contaminated by industrial emissions.

Another dimension of the modern industrial food chain that has so far escaped close scrutiny is the impact of widespread use of antibiotic chemicals in livestock-raising. Worldwide, there is a growing pandemic of deaths and serious illnesses caused by the fact that certain bacteria are becoming partly or totally resistant to modern antibiotic drugs (see Figure 3). The US CDC estimates that at least 23,000 Americans die annually due to infections caused by resistant bacteria, and many thousands more from the resulting complications.[29] In Europe, the death toll is estimated at 25,000 a year.[30]

The CDC *Threat Report* makes it plain that part of the problem originates with modern livestock production systems. 'Drug-resistant bacteria can remain on meat from animals. When not handled or cooked properly, the bacteria can spread to humans,' says the CDC. 'Fertilizer or water containing animal feces and drug-resistant bacteria is used on food crops. Drug-resistant bacteria in the animal feces can remain on crops and be eaten.

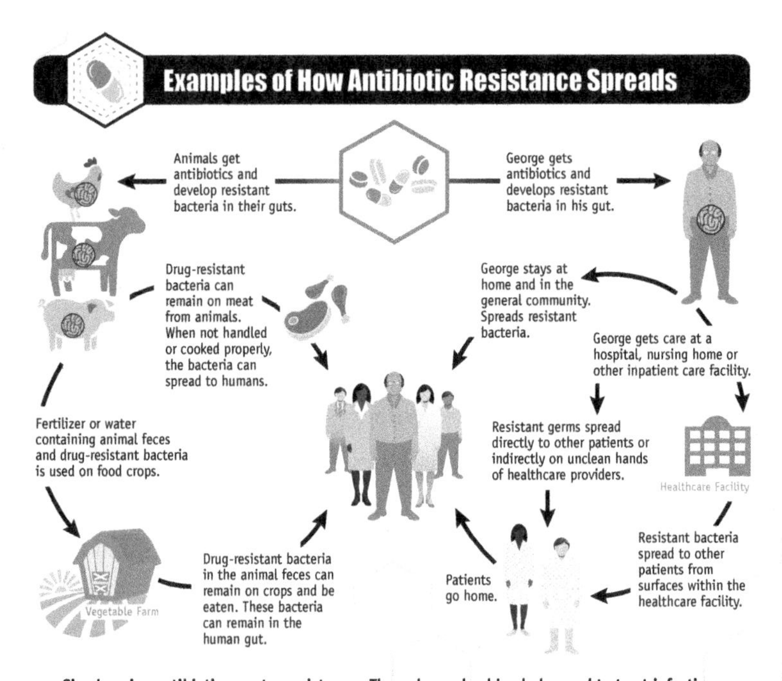

FIGURE 3 How antibiotic resistance spreads

Source: Centers for Disease Control (CDC), 'Antiobiotic resistance threats in the United States', 2013, p. 14.

These bacteria can remain in the human gut.' Since the lion's share of antibiotic use in Western countries is in the livestock sector, it is likely that a significant proportion of the human toll from drug-resistant microbes is due to the modern food system, not simply to the overprescribing of antibiotics by doctors. It is a further drawback of the chemicalisation of the food system.

It is important to remember that this almost universal penetration of man-made chemicals into the food chain has mainly happened in just the last half-century, particularly in the last twenty-five years. No previous generations were so exposed—although people would have faced naturally occurring toxins, as some plants, for example, take up certain heavy metals from soil, and many secrete natural poisons to prevent destruction by animals and insects. However, the toxic exposure of humans through their diet for thousands of generations would generally have been quite small as the human diet itself was far more diverse than today. Humanity is now moving into totally uncharted waters; we are exposing ourselves to many thousands of novel substances which, in most cases, do not occur in nature and to which our bodies are therefore quite unaccustomed and have few specific defences against.

Among many consumer groups now attempting to grapple with the expanding octopus of a globally contaminated food supply and its effects on themselves and their families, the Food Intolerance Network (FIN), set up by Sue and Howard Dengate in Australia and New Zealand, provides independent scientific information to consumers and parents about chemicals in food (natural as well as man-made), and the diseases and allergies now linked to them.[31]

As with so many modern families, concern about their diet

began with parenthood and seeking the best for their children's health. Sue, a former teacher, recounts:

> After the birth of our daughter, we felt as if we had been launched into a nightmare world of severe sleep disturbance, temper outbursts, arguments, oppositional defiance, asthma and other health problems that lasted for over ten years. I had always assumed that I would know if my child was affected by food additives, because I would see a reaction soon after eating—but that's not how it works.
>
> Food reactions are usually delayed by hours or even days and are almost impossible to identify without an elimination diet. I regard myself as a trained observer but I couldn't work out exactly which foods were affecting my kids. However, we never stopped looking for a solution to our problems and after ten years of searching, a dietitian offered us a three week trial of the Royal Prince Alfred Hospital Elimination Diet[32] which turned out to be the magic answer for us.

The RPAH diet avoids additives and is also low in salicylates, amines and glutamates. These naturally occurring food components can cause the same problems as additives in some people, when found within whole foods such as some fruits, and especially when concentrated by processing—for example in highly flavoured products such as strawberry yoghurt. The diet ensures all potential dietary problems are removed at the same time. Sue continues:

> We did the RPAH elimination diet as a family, to support our daughter—but we were surprised to find that we *all* improved. The challenges afterwards showed that we were all affected by some additives and natural food chemicals although often in very different

> ways. Between the four of us we experienced every adverse food reaction listed by our food regulators—from rashes and swelling of the skin, asthma and stuffy or runny nose, to irritable bowel symptoms, colic, bloating, diarrhoea, migraines, headaches, lethargy and irritability—and many more. Although we need to live on a restricted diet while at home in Australia, we have been surprised to find that when we go trekking in Nepal, we can eat traditional food in remote Himalayan villages without suffering from the health, learning or behavioural disorders that the Western diet causes us.

Sue's partner Howard Dengate is a food scientist by profession—and very well placed to understand the issues:

> It is the craving of Western supermarkets for 'immortal food', combined with a flawed food additive approval process, that has caused many of the huge range of problems reported by the 10,000-member Food Intolerance Network. Today we receive thousands of emails from people all over the world who have found, like us, that changing their diet has a profound effect and want to thank us for making this information available. Among the many success stories we receive are those from people whose children are now happy, healthy and doing well in their journey through life—and from adults who have battled with chronic conditions such as itchy rashes or painful heart symptoms for years, only to find a change of diet is their answer.[33]

On their website fedup.com.au, the Dengates explain:

> Our food has changed drastically over the last 30 years, and so have food-related problems. Additives are now used in healthy

> foods such as bread, butter, yoghurt, juice or muesli bars as well as in junk food. Reactions to food additives are related to dose, so the more you eat, the more likely you are to be affected. A British survey in 2007 found that most consumers underestimate how many additives they eat, the average consumer eats twenty additives per day (nineteen if foods are home-cooked) and most consumers don't know which foods contain additives.

Besides discussing chemicals and their effects, the Food Intolerance Network provides consumers with a shopping guide to minimise their exposure to chemical additives.

While membership of the FIN is comparatively small on a global scale—several thousand families who are attempting to manage food related ill-health, mainly in children—it nevertheless represents the bow-wave of a growing international movement by consumers to take greater interest in, and responsibility for, their own diets. It signals unmistakably that the educated global consumers of the twenty-first century are joining hands and sharing experiences around the globe, and they are no longer willing to swallow processed foods containing anonymous and potentially toxic substances. Nor are they willing to take on trust the assurances of regulators that food is being properly scrutinised and checked. This is a very important and heartening development, to which I will return in Chapter 9.

However, despite its high profile in the media as a source of concern, food represents but a portion of our total exposure to man-made chemicals and pollutants. Scientific studies which quantify the full extent and effects of our exposure to chemicals from *all* sources are still critically lacking, although as we have seen, some moves are now afoot to remedy this. But clear answers may still be years, perhaps decades, away.

Dubious drinks

Drinking water has one of the highest profiles of any chemical contamination issue in the world—and is the focus of major clean-up and prevention efforts by governments, such as America's drinking water strategy which aims 'to strengthen public health prevention from contaminants in drinking water'. The US Environment Protection Agency (US EPA) lists around ninety chemicals or categories of substance with permitted levels in tap water of either zero, or near zero, on the grounds of toxicity. This list, or variants of it, is widely followed by governments around the globe.[34] Similarly, the World Health Organization lists guideline values for the maximum safe amount of certain chemicals allowable in drinking water.

However, with around 140,000 deliberately made compounds, tens of thousands of adventitious substances, and hundreds of new chemicals being developed and released every year, it is plain that the enormity of the task of keeping the world's water free of dangerous chemicals is ballooning, not receding. The dilemma is particularly acute in the case of newly industrialising giant, China, where a third of the nation's waste water from industry and farming—and more than 90 per cent of China's household sewage—is released untreated into rivers and lakes.

So great is the water pollution problem in China that the World Bank has warned of 'catastrophic consequences for future generations'.[35] Half of China's population lacks safe drinking water. More than 980 million people are estimated to use polluted water for drinking and washing. Water pollution accounts for half of the $69 billion that the Chinese economy loses to pollution every year.

Around 5300 tonnes of organic pollutants are released into China's waters each day, rendering a quarter of its lakes, rivers and streams undrinkable. Big water problems are by no means confined to China—the pollution is being exported around the world, firstly through discharge into the oceans of China's rivers and secondly via the manufacture of contaminated goods, including and especially, food.[36]

Similar problems are evident in another emerging industrial giant, India, where they have been linked to stunted growth in Indian children, according to UNICEF.[37] And the European Environment Agency says 10 per cent of Europe's surface water bodies are 'in poor chemical status'—meaning they are contaminated by harmful compounds or heavy metals—adding that no data is available for 40 per cent of Europe's surface waters. 'Ground water bodies are in a worse condition—approximately 25 per cent of groundwater, by area, has poor chemical status across Europe. Sixteen member states have more than 10 per cent of groundwater bodies in poor chemical status,' it notes, instancing Luxembourg, the Czech Republic, Belgium and Malta.[38] Altogether, it is estimated that bad water costs around 5.1 million lives per year, and is the largest single cause of death—though much of this mortality rate is associated with infection, not due to chemicals alone.

Even clean tap water carries health risks. Many countries add chlorine to their water to kill bacteria, and the chlorine reacts with substances naturally present in the water to produce a host of chemical by-products known as CDPs, many of which have been found to be toxic and some of which are implicated in miscarriages, stillbirths and birth defects. Around thirty countries (representing a combined population of 400 million) either add fluoride to their drinking water to prevent tooth decay or have it naturally present in their water supply. Just what effect all this low-level chemical exposure (and mixing of chemicals) is having on the population is not yet clear—but it augments all the other exposures to which we are subject.[39]

Avoiding tap water in favour of manufactured drinks does not preclude the problem, as these drinks have chemical additives. 'Soft' drinks are not so soft, claims Australian anti-allergy consumer group, Looking for Alternatives. As displayed on the label, a typical fizzy drink may claim the following harmless-sounding ingredients: carbonated water, sugar, colour, food acid, flavour and caffeine. Converted to disease potential however, these ingredients have been separately and collectively linked to tooth decay, obesity, neurotoxicity, hyperactivity, kidney and gut problems, skin and eye diseases, cancers and gene damage.[40] For those trying to escape the dreaded sugar by inclining towards so-called 'diet drinks', the potential risks are similar if not greater, depending on the chemicals used in them: among the risks are cancer, heart disease, asthma, clinical depression, migraines, insomnia, diabetes, blindness, palpitations, hyperactivity and seizures. This is not to say that a specific drink will cause any or all of these diseases but that, together with all the other risky chemicals in our food, air, water and surroundings, they increase our total exposure and add to the likelihood of an unfortunate outcome. It is this cumulative exposure, about which we know so little, that is the real concern.

Those apparently harmless beverages enjoyed daily by many people—coffee and tea—also contain their fair share of toxins—typically about a dozen different ones per cup, according to Food Standards Australia and New Zealand, a food regulation and testing agency that tested eight different kinds of coffee purchased in Sydney and Melbourne, for 133 different contaminants. These included metals and metalloids (arsenic, aluminium, copper, tin, nickel) mainly taken up from the soil during coffee bush growth, and acrylamide—a by-product of the roasting process.[41] The good news is that the levels were very low—too low to cause immediate health issues, the researchers said. But nevertheless, since humans tend to accumulate heavy metals in our bodies, coffee and tea make their own small, yet refreshing, contribution to a more polluted you.

Ageing accumulators

The rate of chemical uptake by the human body may increase with age, recent studies suggest. This is because older skin is thinner and more permeable to toxins, and the kidneys, liver and blood supply are less efficient at breaking them down and removing them from the body.[42]

Nowadays, too, there is a trend towards obesity in older people, and fat is where many persistent substances (such as POPs) accumulate. The longer such substances remain in the body and the more they concentrate, the greater the likelihood of health damage. This suggests that the incidence of chemical poisoning and related diseases is likely to rise as an additional unanticipated healthcare burden of the ageing, obesifying population. Accumulated toxins in older people may also interact with other diseases or effects of ageing, leading to a weakened immune system, which may increase the risk of cancer for example.

Post-mortem pollution

Death itself marks no boundary to the processes by which toxic substances may be released and recirculated. 'Almost all cemeteries have some potential for pollution,' comments a world-first study of the emissions from burial grounds, carried out in Australia.[43] This found that most of the long-lived chemical substances contained within a corpse at the time of death are re-released into groundwater within ten years of its burial, thus exposing future generations to accumulated toxins from the present age. While the most obviously risky discharges from burials are bacteria and viruses, there are reports of more subtle and long-lasting releases of heavy metals and persistent organic pollutants being found in groundwater downstream of burial grounds—groundwater which, not infrequently, ends up as urban tap water. Humans, as long-lived, top bioaccumulators, store more of these substances in their bones and fat than do other animals—and give them back to future generations in a process which has all the hallmarks of a Pharaonic curse.

While cemeteries are not to be compared with, say, landfills or contaminated industrial sites as a concentrated source of future health risks, these findings nevertheless illustrate the long toxic life of some metals and compounds and their capacity to recycle within the Earth system and human population, potentially entering future water supplies and food chains. This raises disturbing reflections that chemical toxicity may now, in some senses, accumulate generation upon generation, a morbid heirloom. Its impacts may reach into the distant future and to our descendants yet unborn.

The toxic toll

The combined human death and disease toll from man-made chemicals is not known, but WHO conservatively estimated in 2012 that 4.9 million annual deaths (8.3 per cent of all deaths) and eighty-six million disabilities per year are attributable to man-made contamination in one form or another.[44] Of these, one million deaths and twenty-one million disabilities are directly attributed to acute cases of chemical poisoning. WHO adds 'This global estimate is an underestimate of the real burden attributable to chemicals.' For example, it does *not* include millions of deaths from cancer (now the world's leading killer), heart disease, obesity, diabetes, stroke, lung disease, genetic disorders and mental disorders, in which chemicals are increasingly implicated by scientific research (see Chapter 6).

Nevertheless, even at this low estimate, contaminants now rank among WHO's 'top ten' leading causes of human death, claiming seven times as many lives as malaria, four times as many lives as road crashes and more than double the toll from HIV, for example.[45] The irony is that this afflicts mainly developed and rapidly developing societies: chemotoxicity is a disease of prosperity.

The Blacksmith Institute, an international not-for-profit organisation that, with Green Cross Switzerland, publishes an annual survey of the world's most polluted sites, estimates that 'close to 125 million people are at risk from industrial pollution worldwide'.[46] However, even this ground-breaking report focuses chiefly on contaminated sites, industrial areas and their immediate surrounds, mainly in poor and medium-income countries, and not on the far larger and more diffuse issue of Earth system contamination from all sources, including food and drink. The report therefore represents a very conservative estimate both in terms of scale and pervasiveness of the true impact of human chemical emissions.

Indeed it is probable that, out of around eight billion people on our planet, nobody is unaffected by our universal chemical suffusion (see Table 2). We each have a compelling reason to do something about it.

TABLE 2 World's main toxic pollution sources, by rank

1.	Battery recycling
2.	Lead smelting
3.	Mining and ore refining
4.	Tanneries
5.	Industrial/municipal dumps
6.	Industrial estates
7.	Artisanal gold mining
8.	Product manufacturing
9.	Chemical production and use
10.	Dye production and use
11.	Petrochemical processing
12.	Electronic waste
13.	Heavy industry
14.	Pesticide production and use
15.	Uranium processing

Source: Blacksmith Institute, 2012.

CHAPTER 4
DIABOLIC COCKTAIL

Seemingly insignificant quantities of individual chemicals can have a major cumulative effect.

Colborn, Dumanowski and Myers, *Our Stolen Future*, 1996

For eight days in 2010, five San Francisco families were tested by researchers for the presence of common hormone-disrupting chemicals used in food packaging. The families ate their normal diet for five days, and a healthy fresh food diet for the remaining three, then had their urine tested. The scientists concluded the family members had ingested a cocktail of hormone disruptors during the normal diet phase, which declined dramatically when they moved onto fresh food.[1] *The Washington Post*, in its report on the research, commented, 'scientists are beginning to piece together data about the ubiquity of chemicals in the food supply and the cumulative impact of chemicals at minute doses. What they're finding has some health advocates worried.'[2] It went on to quote policy director Janet Nudelman at the non-profit Breast Cancer Fund as saying this was 'a huge issue, and no [regulator] is paying attention. It doesn't make sense to regulate the safety of food and then put the food in an unsafe package.' The report added that the US Food and Drug Administration had approved more than 3000 of these chemicals for use in food-contact applications since 1958.

Further to the issue of toxic pesticides making it to the meal table described earlier, the European Pesticide Residue Survey explains 'consumers are expected to be exposed to multiple (pesticide) residues present on food eaten with one meal, during one day or over a longer period which may lead to cumulative (additive or synergistic) effects on human health.'[3]

Chemicals used in the packaging, storage, and processing of foodstuffs may also harm human health over the long term, a team of environmental scientists warned in a commentary in the *Journal of Epidemiology and Community Health*. This is because most of these substances are not inert and can leach into the foods we eat. People who eat packaged or processed foods are likely to be chronically exposed to low levels of these substances throughout their lives, they say.[4] Scientists have long known that air pollution causes various health problems including heart disease—but constant exposure to an estimated 4000 known 'food contact materials' (FCMs) in food packaging is equally common for most people in modern society. Dr Jane Muncke of the Food Packaging Forum Foundation, Zurich, says: 'FCMs are a significant source of chemical food contamination, although legally they are not considered as contaminants. Current testing methods do not cover the risks which this low-level, lifelong exposure may cause . . . It is a major challenge—for epidemiology, toxicology and other health and life sciences—to tease out the cause–effect relationships between food contact chemicals and chronic diseases like cancer, obesity, diabetes and neurological and inflammatory disorders.' Yet this is essential to the prevention of chronic disease, they conclude.[5]

All this highlights a sleeping giant of an issue—that humans are now being affected not just by individual chemicals of concern, but by an enormous cocktail of chemicals from numerous apparently innocuous sources in our daily lives, which may act in concert with one another and affect our health in ways we do not as yet understand.

Whether it is a blend of toxins and heavy metals leaching from a former industrial site and getting into our air, water or food, hydrocarbons entering groundwater from a former petrol station or fuel dump, a witches' brew of poisonous substances from an old urban landfill dissolved in our drinking water, a cocktail of pesticides, preservatives, chemical dyes and additives in the daily diet or a whiff of volatiles from cosmetics, furnishings, clothing and vehicles, modern humans are constantly assailed by hundreds and probably thousands of different man-made chemicals every day—indeed almost every moment—of their lives, in ways that previous generations simply were not. Despite the magnitude of this assault—and a growing global scientific research effort led by the US, Europe and Japan to understand and measure it—the health effects of these chemicals, in both the short and long term, remain largely unquantified and unknown. Many scientists and doctors, however, are now gathering evidence for its impact on the emergence of entirely novel, or previously rare, ailments (as we will see in Chapter 6).

'Real-life exposures are rarely limited to a single chemical and very little information is available on the health and environmental effects of chemical mixtures,' is how the UN Environment Programme (UNEP) describes the problem.[6]

One case in which a chemical mixture was shown to cause actual harm was an experiment by Danish scientists in which lab rats suffered deformed penises and a variety of other gender problems when exposed to just four of the many substances common in human food and packing and which interfere with male hormones. They reported:

> Strikingly, the effect of combined exposure to the selected chemicals on malformations of external sex organs was synergistic, and the observed responses were greater than would be predicted from the toxicities of the individual chemicals. In relation to other hallmarks of disrupted male sexual development, including changes in AGD (ano-genital distance), retained nipples, and sex organ weights, the combined effects were dose additive. When the four chemicals were combined at doses equal to no observed adverse effect levels estimated for nipple retention, significant reductions in AGD were observed in male offspring.[7]

The scientists concluded that 'because unhindered androgen action is essential for human male development in fetal life, these findings are highly relevant to human risk assessment. Evaluations that ignore the possibility of combination effects may lead to considerable underestimations of risks associated with exposures to chemicals that disrupt male sexual differentiation.' The study was significant in other ways. Crucially, it showed that the effect of a chemical mixture ingested by a mother can pass to her offspring. It has subsequently been demonstrated that the impact of chemical exposure in a great-grandmother can be transmitted to her descendants over several generations.[8]

Also, if the rat trial has anything like the implications for humans that its authors suggest, it implies that future generations of human males may not only be less well endowed than their forebears, but also suffer genital deformities. All because of something their mother ate, which is now commonplace throughout the modern industrial food chain. At the very least, this conjures up the spectre of an immense evolutionary chemistry experiment being practised upon entire populations without their knowledge or consent: most of us have no more choice than the lab rats did.

Mixture troubles

'Humans and all other organisms are typically exposed to multi-component chemical mixtures, present in the surrounding environmental media (water, air, soil), in food or in consumer products,' wrote Professor Andreas Kortenkamp and co-authors in the EU's *2009 State of the Art Report on Mixture Toxicity*. 'However, with a few exceptions, chemical risk assessment considers the effects of single substances in isolation, an approach that is only justified if the exposure to mixtures does not bear the risk of an increased toxicity.'[9] The report points out chemical mixtures can arise in several ways:

- as a result of substances in which several chemicals are already mixed

- in manufactured products combining several different chemicals
- through interaction during production, processing, transport, consumption or recycling
- by chemicals becoming mingled in air, water, soil, plants and animals, food and in human tissues, via many different pathways.

To put it plainly, modern chemical risk assessment, however well intentioned and thorough, tends to sidestep the big picture, especially at the global scale. As David Carpenter and colleagues from New York State University put it: 'In reality, most persons are exposed to many chemicals, not just one or two, and therefore the effects of a chemical mixture are extremely complex and may differ for each mixture depending on the chemical composition. This complexity is a major reason why mixtures have not been well studied.'[10]

Add to this the fact that even many individual components have also only been poorly studied: 'Of the tens of thousands of chemicals on the market, only a fraction has been thoroughly evaluated to determine their effects on human health and the environment,' cautions the UNEP. Even in the best-regulated societies, chemical toxicity is habitually investigated one chemical at a time; no overall true impression of the nature or scale of the risks we run from simultaneous and long-running exposure to chemicals from all sources combined is available.

Although many mixture-toxicity experiments have been carried out in the laboratory—chiefly by exposing simple organisms such as worms, plant seeds and microbes to carefully chosen mixtures—most of these have so far presumed that the increase in toxic risk is arithmetic, adding the risk factor for one chemical to that of another to obtain a total estimate for the mixture as a whole. However, recent research suggests the risks when two or more chemicals are present at the same time may instead be non-linear, meaning that different chemicals not only add toxicity on their own behalf but may also exacerbate the toxic impact of others or else react with other chemicals in the mixture to form new compounds which could be overall even more poisonous. Underlining this, the authors of the EU mixture toxicity report state:

> There is strong evidence that chemicals with common specific modes of action work together to produce combination effects that are larger than the effects of each mixture component applied singly. Fewer studies have been conducted with mixtures composed of chemicals with diverse modes of action, with results clearly pointing in the same direction: the effects of such mixtures are also higher than those of the individual components.
>
> There is a consensus in the field of mixture toxicology that the customary chemical-by-chemical approach to risk assessment might be too simplistic. It is in danger of underestimating the risk of chemicals to human health and to the environment.

Until recently, too, there was a presumption that if every chemical of concern in a mixture were to be kept below its particular threshold of known harm, then the mixture overall would be safe. Tests in microbes, water fleas, fish and human cells have cast grave doubt on this idea, the EU mixture report states. 'Hence, any concentration of any compound needs to be considered because it adds to the mixture concentration.'

Furthermore, even in cases where the various chemicals in the mix have completely different toxic actions, the potential harms they pose still cannot be considered in isolation, as they may interact with one another, or form compounds that then affect the consumer's body in unexpected ways. 'There is decisive evidence that mixtures composed of chemicals with diverse modes of action also exhibit mixture effects when each component is present at doses equal to, or below points of departure (harm thresholds),' the European scientists affirm. 'Whether or not risks arise from combined exposures can only be decided on the basis of better information about relevant combined exposures of human populations and wild life. This information is currently missing, and this presents a major challenge to risk assessment.'

Subsequently, the EU Directorate-General for Health and Consumers has said: 'Since humans and their environments are exposed to a wide variety of substances, there is increasing concern in the general public about the potential adverse effects of the interactions between those substances when present simultaneously in a mixture . . . In view of the almost infinite number of possible combinations of chemicals to which humans and environmental species are exposed, some form of initial filter to allow a focus on mixtures of potential concern is necessary . . . A major knowledge gap at the present time is the lack of exposure information and the rather limited number of chemicals for which there is sufficient information on their mode of action.'[11]

In short, as the European health authorities caution, we simply do not clearly comprehend the danger we may be in, or the risks we continue to run as members of a complex industrial consumer society. The information that could help us to make safe and responsible choices is currently at best highly fragmentary—and it tends to suggest that by avoiding particular products, or by changing our behaviour in certain ways, we can eliminate or mitigate the risk to ourselves. However the emerging evidence indicates this may be a delusion. Only by reducing the total number, volume and overall toxicity of *all* the man-made chemicals present in our daily lives can we limit or reduce the risk. But there is little indication that this is about to happen, as Chapter 5 will show.

Billions of mixtures

Given the many billions of possible mixtures that the 140,000 man-made chemicals can form—not to mention the tens of thousands of other adventitious substances created or found in society's waste streams, mineral discharges and fossil fuel emissions they can interact with—the task of assessing every possible mixture for its effect on human health might well appear insuperable. This appears to be the main pretext as to why such assessments are not being attempted by governments or industry on any meaningful scale. However the authors of the EU report consider this task is not impossible: 'There is strong evidence that it is possible to predict the toxicity of chemical mixtures with reasonable accuracy and precision. There is no need for the experimental testing of each and every conceivable mixture, which would indeed make risk assessment unmanageable,' they say. This applies especially to mixtures of known chemical composition (for example, most mixtures found in our food) where both dose rates and the independent action of substances in the mix can be considered together. Where ingredients are incompletely known (for example, as in the case of an old industrial sludge), the problem remains challenging.

However, with the health and maybe even the lives of tens of millions of people at stake, there is no justification for ignoring this challenge on the mere grounds that it is difficult or expensive. The advent of sophisticated chemical and genetic modelling, and supercomputers which allow researchers to quickly perform vast numbers of calculations, make a reasonably educated prediction about the mixtures that pose the greatest potential hazard, and the groups of people who are most at risk. At Australia's CSIRO, for example, Dr Amanda Barnard and her team are using supercomputers to predict the health impact of novel nanochemicals with precisely this intention—to avoid the problems of toxicity and contamination that have accompanied the rise of conventional chemistry.[12] Similar research is being carried out in the US and Europe.

Over a lifetime, each of us is exposed to highly complex mixtures of many thousands of chemicals, generally in small doses, sometimes for short periods, sometimes constantly over many years and sometimes in large pulses. Some of these substances we flush from our bodies readily, from some we sustain transient damage and some (such as certain heavy metals and persistent chemicals) we bioaccumulate over many years, building up our own potentially toxic dose over repeated exposures to a multitude of non-toxic doses. However, the extent to which chemical mixtures attack or undermine our health, and that of the environment, over time remain scientific unknowns.

America's CDC has a Toxic Mixtures research program that is 'mandated to determine the health impact of exposure to combinations of chemicals', but which makes the immediate concession that its main focus is on those mixtures most likely to be found at hazardous waste sites. These, as we have seen, represent only a fraction of the possible sources of chemical mixtures in the average person's daily life and the approach conforms to the out-dated notion that contamination is a local, rather than a universal, issue. This is not the fault of CDC—rather it reflects the low priority which governments and society still attach to the issue and the resulting inadequacy of research funding to investigate it properly.

The exceptional complexity of this challenge has led organisations such as Australia's Cooperative Research Centre for Contamination Assessment and Remediation of the Environment (CRC CARE) and many others to investigate ways to reveal the overall state of toxification of an individual *from all sources*, rather than attempting to separate and attribute the individual's ailments to particular substances. This may employ a range of indicators, including levels of certain proteins in the blood, immune system status, measures of genetic damage and so on. Although it is imprecise in tracking down the specific poisons at fault, this approach—if successful—at least has the virtue of disclosing how poisoned an individual may be in total (at a particular time and from multiple causes). If it is used for general screening and combined with epidemiological studies, it may then be possible to narrow down the main sources of toxicity and identify which ones are doing the most harm. Also, the study of many different individuals may enable a society-wide estimate which may reveal the broad patterns of toxification in various groups of exposed people and hence, the likely main sources of their particular exposure.

Mixtures and health

The effect of chemical mixtures on human health is, as might be expected, as complex and murky as the mixtures themselves. Many chemicals appear to act on most of the cell types in our body. They may invade the cell and scramble its functions, they may reprogram what its genes do, causing it to affect other cells, or they may kill it. They may simply block the cell's ability to receive the normal 'healthy' chemicals we rely on for life, such as calcium.

Where there are several chemicals at work, they can have a wide range of different and interacting effects, making accurate diagnosis especially problematic for doctors grappling with a range of non-specific symptoms. For example, each chemical may have quite different actions on the kidney, the liver and the brain, each with a different disease effect—sometimes even a benign one—or else there may be a confusing combination of benign and malignant effects.

A celebrated study by the World Health Organization in 1996 estimated that 80 per cent of all cancers were due to 'environmental factors': exposure to things in our daily living environment such as chemicals, radiation and tobacco smoke.[13] A team from the University of California at Berkeley then estimated that roughly 25 to 33 per cent of the worldwide burden of disease was linked to 'environmental factors'.[14]

It is an unfortunate illustration of the propensity of science to confuse the public by referring to man-made substances as 'environmental factors', as if these were naturally occurring things. In many cases the phrase definitely means 'things put into the human living environment by humans', not things which occur naturally in the natural environment. It is worth reflecting on how a person suffering one of the many chemical-caused diseases, when told by their doctor that it is due to 'environmental factors', might assume the cause to be natural—and accept as 'fate' a disease that is in fact due to preventable or avoidable human activity. The use of such ambiguous and misleading terminology thus helps to conceal from the public the extent of the risk from chemicals in general and chemical mixtures especially.

In some cases, indeed, society may be deliberately misled by the use of such language, by industry or the scientists who work for it and wish to divert attention away from industrial pollution. The effect of widespread public ignorance about the scale of the threat is to reduce political pressure for it to be remedied. This in turn results in a low priority—and lower funding—for research specifically intended to explore the impact of chemical mixtures. Thus, by its poor choice of language at the critical interface between science and society, science itself may be in some cases unwittingly ensuring it does not have enough funding to properly investigate what is plainly a life-and-death issue affecting the whole of humanity.

Nevertheless, a growing number of pioneering studies by scientists have revealed that a wide range of diseases not previously thought to have been caused by 'environmental factors'—such as heart disease, diabetes, bone and joint disorders—are in fact linked to exposure to man-made contaminant mixtures. Precisely what the link was, or how much it is influenced by mixtures of chemicals as opposed to single substances, remains hard to define: 'In many of these situations we have some evidence for a variety of chemicals having a relationship to the disease, but we currently have little or no evidence of the nature and effects of interactions among chemicals related to the disease state,' explained David Carpenter and colleagues from New York State University (NYSU).[15] They divided the chemicals studied into two types: those that interfere with normal development and cell function, and those that cause direct cell damage.

The NYSU researchers identified the following broad categories of human disease as being partly derived from the effect of chemical mixtures:

Developmental disorders: disorders that occur when exposure to chemical mixtures in pregnancy, infancy or childhood leads to lifelong mental, physical and reproductive disorders.

Neurobehavioural abnormalities: including a lifelong deficit in intelligence caused by exposure to toxins such as lead and PCBs in the early years; nerve disorders brought on by mercury exposure; intelligence and coordination problems due to pesticide exposure; reduced learning capacity caused by mixtures containing PCBs, dioxins, furans, lead and mercury.

Sexual disruption: caused by chemicals which mimic or alter the natural functioning of sex hormones in the very young. These result in feminisation of males; masculinisation of females; reproductive deformities; diseases of the womb and prostate; infertility in both males and females; reproductive cancers; and possibly changes in sexual preference.

Neurodegeneration: includes a range of diseases that result in the death of nerve and brain cells, such as Parkinson's, Alzheimer's and amyotrophic lateral sclerosis (ALS). While the links are still largely speculative owing to the lapse of time, more scientists are now attributing these mature-onset diseases to a person's early exposure to toxic chemical mixtures.

Cancers: genetic factors alone are thought by scientists to be responsible for no more than 5 per cent of cancers. The rest have an external trigger, although there may be a genetic predisposition that makes the individual more susceptible to the particular cancer. The biggest 'killer mixture' of chemicals is delivered in cigarette smoke (accounting for 30 per cent of all cancer deaths), but there are plenty of other suspect chemicals, including oestrogens, organochlorine pesticides, PCBs and polyaromatic hydrocarbons, which have all, for example, been linked with breast cancer.

Heart disease: while heart disease and stroke are most commonly linked to fat and diet, they are also quite strongly connected in the scientific literature with exposure to chemical mixtures, containing for example heavy metals, certain pesticides and cigarette smoke. The theory is that these chemicals not only damage cell function, but also cause direct cell damage.

How chemical mixtures actually cause these diseases is a matter of extraordinary biochemical complexity, which is still in the process of being unravelled by science. The slow speed of this meticulous process is partly responsible for the low level of general awareness—in society as a whole, and among governments in particular—about the complex risks posed by chemical mixtures. It is important to emphasise that this does *not* mean the risks are small—only that they are extremely difficult to pin down and quantify. In other words we are probably underestimating the dangers.

Humans in today's world are subjected to an unprecedented bath of man-made chemical mixtures head-to-toe, every day of their life. While some scientists argue that the fact that more people are alive and living longer bespeaks our ability to adapt to the toxic flood, others are not so sure, fearing that worse may be in store for the health of both individuals and society as a whole in coming years, unless the weight of our overall chemical burden is somehow reduced.

At present there is no practical way to contain the impact of chemical mixtures—either on humans or on the natural environment. Since these substances are emitted from multiple sources—not only the chemical industry itself but also mining, power generation, manufacturing, transport, construction, pharmaceuticals, waste disposal, food, cosmetics and so on—and reach people by multiple pathways, it is almost impossible to regulate or control their production, dispersal and recombination in mixtures. And if the pattern established by tobacco multinationals is anything to go by, an industry which deliberately or unintentionally produces toxic substances will as a general rule begin by denying its contribution, and then blame every other industry for the impact of mixtures before perhaps at some point very reluctantly acknowledging its own contribution, and agreeing to curtail it. This process usually spans decades.

The task for government in pursuing industries whose products may in some cases only be harmful when interacting with other chemicals in mixtures is even harder, as the trail of responsibility is more difficult to trace and assign. Yet it is probable that millions of people are suffering and dying—and more will die in coming decades—as a consequence of the general release of these substances, their combined interaction and our reluctance or inability to investigate the true scale of the risks we run.

For millions of people to die needlessly for any *preventable* reason is an affront to civilisation. It is a basic moral principle that such preventable deaths should not be tolerated by society any more than other forms of manslaughter. Yet, in the case of chemicals, we have somehow forsaken our duty of care—or chosen to overlook it. The millions of infants and children who will have their young lives affected by exposure to these noxious mixtures before even attaining an age where they can refuse, reject, or seek to avoid them represent a grievous delinquency on the part of society. We lust for the goods and benefits of modern consumerism. Yet we turn aside from the moral responsibility for its impact.

CHAPTER 5
UNSEEN RISKS

I wasted time, and now doth time waste me.

William Shakespeare, *Richard II*

Amid the acrid reek and choking smog from open fires and smoking iron cauldrons, in a place where it seemed half the world's computer scrap was being rendered down just to recover a few miserable metal traces, a young Chinese scientist was studiously gleaning information to help her team understand the health problems now starting to emerge among the workers and children enslaved in this New Age Inferno. As she spoke with some of them, a group of rough-looking men approached, demanding to know what she was doing. When she tried to explain her intention of gathering information for her team's health research, the knives came out. One of the men slashed at her, opening an ugly wound on her hand with his knife, as she fought to deflect the blow. Bleeding and terrified, she and her colleague fled headlong down grim shanty streets, past hovels and yards heaped with high-tech detritus, still pursued by their angry assailants.[1] It was just another grimy, vicious day in Guiyu, south China, a place that has come to symbolise the ugly face of the IT revolution and been dubbed 'the e-waste dump of the world'. Reporter Tim Johnson of *The Seattle Times* wrote:[2]

> The city is a sprawling computer slaughterhouse. Some 60,000 laborers toil here at primitive e-waste recycling—if it can be called that—even as the work imperils their health amid a runoff of toxic metals and acids . . . Computer carcasses line the streets, awaiting dismemberment. Circuit boards and hard drives lie in huge mounds. At thousands of workshops, laborers shred and grind plastic casings into particles, snip cables and pry chips from circuit boards. Workers pass the boards through red-hot kilns or acid baths to dissolve lead, silver and other metals from the digital detritus. The acrid smell of burning solder and melting plastic fills the air.

Guiyu is also a place where the grubby hands of organised crime have found fresh purchase and, like the illicit drug trade, strangers who ask questions about the health or industrial practices of its workers are not welcome. Among the regions beset with this twenty-first-century plague of desperate and perilous gleaning, two are infamous: Guiyu in China and Agbogbloshie in Ghana, where men, women and children labour side by side amid the poisonous stench of open fires to eke out an existence rendering down these products of luxury: '[The] landscape . . . varies from filthy to apocalyptic. In small workshops and yards and in the open countryside workers dismember the detritus of modernisation. Armed mostly with small hand tools they take apart old computers, monitors, printers, video and DVD players, photocopying machines, telephones and phone chargers, music speakers, car batteries and microwave ovens.'[3]

In Ghana, reports *National Geographic*'s Chris Carroll,

> pillars of black smoke begin to rise above the vast Agbogbloshie Market. I follow one plume toward its source, past lettuce and plantain vendors, past stalls of used tires, and through a clanging scrap market where hunched men bash on old alternators and engine blocks. Soon the muddy track is flanked by piles of old TVs, gutted computer cases, and smashed monitors heaped ten feet high. Beyond lies a field of fine ash speckled with glints of amber and green—the sharp broken bits of circuit boards. I can see now that the smoke issues not from one fire, but from many small blazes. Dozens of indistinct figures move among the acrid haze, some stirring flames with sticks, others carrying armfuls of brightly colored computer wire. Most are children.[4]

According to the Blacksmith Institute, there are at least 500 of these digital hells worldwide, mostly in China, Africa and Latin America, each of them the product of the casual disposal of last year's mobile phone, tablet or laptop by an affluent consumer.

New toxics

In addition to the billions of tonnes of emissions already produced by humanity, new categories of planetary contamination are emerging which consist of substances that are largely untested in terms of their effects on human health or the environment, let alone their impact in combination with other chemicals. These new substances—described in detail in the following sections—are being released onto world markets and into the environment at such a rate and so universally that governments are unable to keep up with testing them, let alone regulating or monitoring their impact: the stable door is left wide open and new toxic horses are bolting almost daily. Through the six main global pathways described in Chapter 2—air, water, food, manufactures, trade and parental transfer—these new toxins disperse rapidly around the planet, regardless of where they were originally produced or discarded. In some cases they are cumulative, leading to a continuous ratcheting-up of humanity's toxic burden. This raises the spectre of new and unanticipated 'asbestos and thalidomide tragedies' affecting all societies, not only those with large toxic waste production or processing. As Shakespeare's famous quotation from *Richard II* reminds us, the price of delay rises with each passing day that action is deferred.

E-waste

Over the last quarter-century, the global electronics industry has produced more than five billion personal computers and seven billion mobile phones, several billion televisions and radios, innumerable music players, tablets, games platforms, refrigerators and other electronic devices and whitegoods. Such is the dramatic rate of replacement of old electronic devices with new ones featuring the latest technology that many new devices are rapidly superseded and become redundant—or at least 'unfashionable'—in a matter of months and are then discarded by their owners: the average life span of a computer fell from six years to just two years between

1999 and 2005.[5] Indeed many of these devices are designed to become obsolescent quite quickly in order to encourage consumers to buy upgraded, better-endowed versions—a practice which has the effect of redoubling the amount of e-waste in circulation around the planet.[6] Customer behaviour in the electronics sector, almost more than any other, epitomises the grip which consumerism has acquired on the human psyche, and the resulting mindless release of toxic wastes which accompanies it.

World production of electronic waste ('e-waste') unleashed by this single sector of the world economy alone was estimated by the UN Environment Programme at forty million tonnes in 2008 and by others at fifty million tonnes—amounting to about seven kilos per person.[7] E-waste is especially hazardous, containing heavy metals—some of which are new to Earth in their pure and concentrated state and are thus of unknown impact on our health and the wellbeing of the planet. E-waste also contains many persistent chemicals and suspected or known carcinogens and endocrine disruptors. To give just a brief idea of their contents, devices such as phones and laptops may contain: epoxy resins, fibreglass, PCBs, PFOA, flame retardants, polyvinyl chloride, other additional plastics, lead, tin, copper, gold, silicon, beryllium, carbon, iron, aluminium, cadmium, mercury, thallium, americium, antimony, arsenic, barium, bismuth, boron, cobalt, chromium, europium, gallium, germanium, gold, indium, lithium, manganese, mercury, nickel, niobium, palladium, platinum, rhodium, ruthenium, selenium, silver, tantalum, terbium, thorium, titanium, vanadium and yttrium. A single device may have as many as forty of these substances in its makeup, several of them linked to cancer, various diseases of the lung, kidney and nervous system, reproductive disorders and birth defects. Some 500 different compounds, many of them toxic, are used in the manufacture of electronic goods.

Eminent environmental scientist and toxicologist Professor Ming Hung Wong, from Hong Kong Baptist University, estimates that more than two-thirds of this waste ends up in China where it is roasted in iron pots on crude open fires—or else dissolved in acid—to recover valuable metals. This process causes massive-scale contamination of the local environment (soil, water, air and food crops) with dioxins and furans, leading to heavy body burdens in local people—children and infants especially—of up to a hundred times or more above the World Health Organization's recommended limits.[8] 'These persistent pollutants end up everywhere—the air, the ocean, or leak into soil and groundwater,' Professor Wong explains. 'This problem has been identified in China, the Philippines, Vietnam, Pakistan, Mexico, Brazil and India. It's no longer a problem that is confined to the villages that deal with e-waste, because when water and soil is polluted, everyone is vulnerable to the food products that are exported from these regions.' His comments highlight graphically how contamination can begin in one place where raw materials are extracted, move to another where devices are made, move to a third far away where they are sold, used and discarded, sent to a fourth as scrap, reprocessed in ways that pollute air soil, water and food and then even return to consumers on the other side of the world either in exports of food grown in the polluted region, or else via wind, water, fish and wildlife. It is a fresh example, if yet another were needed, of the universal dissemination of chemical contaminants in our contaminated world—and how even well-run national regulatory systems are now a scant defence.

E-waste is also a toxic curse created in particular by the wealthy, yet with impacts that fall disproportionately on the poor. The scientific study of the health impacts of these postmodern 'dark satanic mills' is still in its early days, but already researchers are reporting higher incidences of oxidative stress, DNA damage and inflammation among the local people, the likely precursors of heart disease and cancers.[9]

A study by the UNEP, *Recycling—From E-waste to Resources*, predicted that by 2020 e-waste from old computers will have jumped five-fold from 2007 levels in India, and by two- to four-fold in South Africa and China; during this same thirteen-year period, the residues from old mobile phones are predicted to increase seven-fold in China and a staggering eighteen-fold in India.[10] With growth in demand for electronic goods rising by around 5 per cent per year, the amount of hazardous waste produced annually per person is likely to increase from seven kilos a head to twice that amount in the 2020s. However, since most of the waste produced in the past quarter century has been improperly disposed of and is largely still in circulation in the Earth system in air, water, soil, groundwater and the food chain, the combined exposure of humanity and all life on Earth to e-waste toxins is probably rising exponentially. The real health impacts are yet to be seen, partly because many of the deleterious physical effects will be difficult to assign to a particular cause and are of a chronic rather than acute nature.

This situation echoes our experience with asbestos, which has been used worldwide as a cheap building and insulation material, from the 1860s up to the present day. The first cases of asbestos-related disease and death were recorded in 1906. Between the

1930s and the 1950s, knowledge of its dangers rose. However, western governments and most of the asbestos industry long tried to keep the public in the dark so they could continue to mine and use this valuable material. By the 1980s growing public health concern led to the end of mining and use of asbestos in developed countries and its progressive removal from buildings and homes; long and costly lawsuits began, as victims sought damages. However, even today when we understand its lethal nature, asbestos continues to be widely used in many developing countries: the global death toll due to asbestos was recently estimated by the WHO at 107,000 per year.[11]

E-waste is unlike asbestos in one sense: the risks have been widely canvassed throughout the world in the media by concerned scientists and green groups, and many plans and proposals have been made for limiting the harm it may be causing. The UNEP, for example, has urged the adoption of centres of excellence in e-waste processing in countries around the world. Most developed countries are now making some effort to deal with more of their waste within their national borders rather than exporting it to the developing world. Progress is slow, however, and the cost of disassembly and metal extraction is high. Meanwhile, the flood of discarded devices swells with each passing week. Above all, in an industry where conspicuous consumption is driven by cheap retail prices, these electronic devices are not designed for quick, easy and safe disassembly and recycling of their intricate components, and particularly not for the separation of their metals: to research and adopt such designs would render them far too costly. Governments are reluctant to hamper a successful industry by mandating such designs. For the time being, at least, it is cheaper to expose billions of people to the health risks posed by the growing global tide of e-contamination. Stated bluntly, this is a problem for which effective technical and regulatory solutions may well exist—but society remains reluctant to enforce them because to do so would involve increasing the cost of some of its favourite toys.

Nanopollution

One of the crucial factors in the triumph of the global electronics sector is miniaturisation and this lesson has clearly been absorbed by the latest boom industry: nanotechnology. Based on the Greek word for a dwarf (nanos), this term refers to devices and substances with components that measure millionths of a millimetre in scale. To give an idea of scale, this is a thousandth to a millionth of the breadth of a human hair or a water droplet in a fog. Such components are so small that their chemical and physical properties change in quite radical ways, enabling many innovative uses—but equally, risk many unexpected and unknown health impacts. A familiar example of the daily application of nanotechnology is the use of tiny titanium dioxide particles in sunscreens and cosmetics, due to their superior ability to block or reflect the sun, thus reducing the amount of skin damage and the 'ageing' it causes.

The nano industry creates chemical and metallic compounds whose molecular size is so small they can easily pass through the human skin, blood-brain barrier, maternal-foetal barrier, lung lining, digestive lining, cell walls and other normal protective biological boundaries of humans and other living organisms, with consequential—but thus far largely unknown—health impacts.

The number of nanoproducts developed and released has soared in recent years. As of 2013, the international Project on Emerging Nanotechnologies (PEN) listed 1317 nanoproducts on the world market, produced by 587 companies in thirty countries. PEN found that these products covered thirty broad consumer categories including food, cooking, cosmetics, children's toys, furnishings, cleaning products, clothing, health and fitness. Owing to industrial secrecy and the uncontrolled release of new products in many countries, this is likely to be a significant underestimate of the actual number of nanoproducts entering the world market, some of which are even found in human food, according to the US Food and Drug Authority. The OECD is now attempting to establish a register. However, the production of nanoparticles is not confined to intentional processes. They are also released when fossil fuels, tobacco or even forests are burned, or when metals are processed. Thus, nanoparticles may already be implicated in a range of common health issues such as lung diseases.

In a seminal review of the issues around nanoscience, Canadian researcher Cristina Buzea and colleagues stated that:

> nanoparticles have the ability to enter . . . and damage living organisms. This ability results primarily from their small size, which allows them to penetrate physiological barriers, and travel within the circulatory systems of a host. While natural processes have produced nanoparticles for eons, modern science has recently learned how to synthesize a bewildering array of artificial materials with structure that is engineered at the atomic scale . . . some

> nanoparticles can penetrate lung or skin barriers and enter the circulatory and lymphatic systems of humans and animals, reaching most bodily tissues and organs, and potentially disrupting cellular processes and causing disease. The toxicity of each of these materials depends greatly, however, upon the particular arrangement of its many atoms. Considering all the possible variations in shape and chemistry of even the smallest nanoparticles, with only tens of atoms, yields a huge number of distinct materials with potentially very different physical and toxicological properties. Asbestos is a good example of a toxic material, its spear-like fibres so slender as to be measured in nanometres, enabling them to penetrate body cells and cause lung cancer and other diseases . . .
>
> Industrial nanoparticle materials today constitute a tiny but significant pollution source that is, so far, literally buried beneath much larger natural sources and nanoparticle pollution incidental to other human activities, particularly automobile exhaust soot.[12]

The researchers caution that the impact of nanotoxicity—the term describing poisoning by nanoparticles—may have been exaggerated in the media, because in many nanoproducts such as paints or electronic devices the particles are fixed in place, reducing the likelihood of health effects: 'While uncontained nanoparticles clearly represent a serious health threat, fixed nanostructured materials, such as thin film coatings, microchip electronics, and many other existing nanoengineered materials, are known to be virtually benign.'

However such views overlook a fundamental fact about nanoparticles: once released into the global environment, most can never be retrieved. Being very small, they are readily transported in the air, in water and in living creatures and this pollution can rapidly disseminate planet-wide. If certain particles turn out to be harmful to life, they can potentially continue causing harm *forever*—unless immobilised or broken down by natural processes. Even nanoparticles which are embedded in paints, metal objects or electronic goods may after a time be released into the environment when those products degrade or are intentionally recycled: today's nano emissions, where they prove harmful, may thus haunt future generations for a long time to come. In a sense, also, every nanoparticle produced adds cumulatively to worldwide nanopollution and while this is still very small relative to other sorts of contamination, it is growing at dramatic rates as hundreds of new products, whose health implications are for the most part largely unknown and unstudied, emerge onto the market. The fact that many of these products are intended for personal use on the skin and in food increases the likelihood of an unfortunate outcome. There is a risk that the story of nanoproducts may thus echo that of tobacco smoking, which has led to numerous deadly diseases never suspected when tobacco was first introduced. To gain an appreciation of these emerging risks, a visit to the website of the International Council on Nanotechnology, at Rice University in the US, reveals hundreds of reports by scientists from all over the world exploring or reporting harmful effects of nanosubstances.[13]

Unfortunately, in common with traditional chemical approaches to the study of global pollution, the main thrust of world science is to treat nanotoxicity as if it were a hazard that is somehow separate from other forms of contamination. Unanswered is the complex and troubling question of how nano substances may interact with other forms of contamination, including e-waste, pesticides, industrial pollution, dioxins, fossil fuel emissions, food additives, chemical mixtures and so on—and how nanosubstances may possibly undermine our ability to withstand harm from these other sources or act synergistically with them to increase the damage we take. Our understanding of this issue is confined to a view of a few individual 'trees', rather than a view of the contamination 'forest' as a whole. And while many of the makers of nanoproducts have professed a desire to make them as safe and healthy as possible and for governments to take the first steps towards regulating them, neither approach as yet encompasses the combined impact of nanopollution with other forms of contamination, affecting everyone every day.

In short, nanotechnology offers a fresh reminder that we humans, and all life with us, are the experimental animals in a vast global chemistry experiment—involving a myriad of substances, some old and some new, some intentional and some accidental—the outcome of which remains unknown, but which our past experience clearly suggests could spell danger for many.

Nutrient discharges

One of the largest of human emissions, in terms of sheer volume, is our release of nutrients into the world's waters. Yet this is also one of the effects that seem to pass largely unnoticed by the general public, as it usually involves the gradual transformation over many years or decades of clear, clean, healthy rivers, lakes and seas into turbid, stagnant and ecologically impoverished—sometimes lifeless—systems.[14] Such very gradual changes have been observed, measured and recorded by science for more than 150 years.

However in recent times more dramatic, sudden changes have also been taking place, where a previously stable environment reaches a 'tipping point' and 'flips'—often permanently—into another state, with the loss of most of its previous plants and animals.[15]

As mentioned in chapters 1 and 2, the primary chemical culprits in this progressive destruction of the world's fresh and marine water bodies are 121 million tonnes of nitrogen and 9.5 million tonnes of phosphorus that are released by human activity worldwide each year—levels very much larger than were circulated naturally around the planet before civilisation began. In the past few decades we have doubled the discharge of nitrogen into the environment and tripled the flow of phosphorus.[16] The main sources of these nutrients are agricultural fertilisers, soil erosion (due to poor farming practices or poorly designed cities, roads and land developments), burning of fossil fuels, and the inadequacy of waste- and sewage-disposal systems. These lost nutrients are causing profound changes in our rivers, lakes and oceans, turning once-productive waters brimful of life into empty, stagnant 'deserts' where few fish and other oxygen-reliant life forms, if any, can survive.[17]

Nutrient pollution is not new. What is clear, and the reason for including it in this chapter as a 'new toxic', is that it is almost certainly bound to double, and possibly triple, by the 2060s—as the world strives to grow enough food to feed a population of ten billion predominantly middle-class citizens. This need for a doubling in food production also implies a possible doubling in the release of fertilisers and a doubling in the release of food waste, including sewage, into the environment. This would have a potentially horrendous impact on all the world's main water bodies—and the life in them. Since nutrients are widely viewed as a benign part of our daily diet, the prospect of a doubling in their worldwide release into the environment has not attracted the same degree of concern from policymakers, media or the public as pollution from other contaminants.

Much has been done, and is being done, to try to reduce nutrient flows at the local level into major rivers, lakes or coastal areas such as Chesapeake Bay in the US, the Baltic Sea around Denmark, the rivers of New Zealand or the Great Barrier Reef Lagoon in Australia. These, however, are simply fresh examples of 'contaminated site syndrome', in which people regard local nutrient pollution as being of concern meanwhile paying little attention to the far larger and more pernicious pollution that is flooding through the Earth system as we shift farming offshore and source our food supplies globally. While 'think global, act local' is sound advice (provided everyone adopts it), dealing with the nutrient issue one bay or lake at a time is a bit like operating on a patient with widely disseminated cancers, one tumour at a time. This surgery may treat the symptoms, but it cannot save the patient.

A doubling or tripling in total human nutrient release will almost certainly result in major tipping points being reached in many water bodies, including large areas of ocean. These in turn will eliminate entire fisheries, prevent fish farming and exacerbate dangerous aspects of climate change. For example, as the oceans stagnate, the microalgae and plankton which absorb carbon and produce oxygen are gradually replaced by anoxic bacteria which produce poisonous hydrogen sulphide gas and kill off other water species. As a result, the oceans absorb less CO^2 and the atmosphere more—and global warming speeds up. This is the process that scientists believe was partly to blame for the worst extinction event in Earth history, at the end of the Permian era, when 90 to 95 per cent of all life forms died out.[18]

So nutrient pollution, while not attracting the same attention as other, more dramatic forms of contamination, nevertheless has potential for a very large impact on the human destiny. Fortunately it also has a feasible technical solution: to close the nutrient cycle and re-use all our wasted nutrients in food production. However this calls for the extensive redesign of both cities and farming systems worldwide, even maybe replacing traditional farming with intensive forms of food production such as algae culture, aquaponics and biocultures, based in cities. Few nations in the world are seriously contempating such a profound shift and the global nutrient flood remains a serious blind spot if we are to live sustainably on this planet.

Novel pesticides

When it comes to contaminants, pesticides have drawn the main glare of public and regulatory attention, with the result that some—especially organochlorines and organophosphates—have been banned in well-run countries (though many developing countries still manufacture them, and they still pose a growing menace to the Earth system and all its inhabitants). Although accurate statistics are hard to come by, owing to the secrecy in which the industry enshrouds itself, the US EPA estimated global use of all pesticides at around 2.37 million tonnes a year in 2007: about *thirty times* larger than it was when Rachel Carson first warned the world about its impact, in the early 1960s.[19] In recent times, pesticide use has been falling slightly by volume in developed countries—as manufacturers replace older broad-spectrum chemicals with more powerful, specific pesticides—and rising in developing countries where both manufacturing and use of the cheaper, more harmful products is growing rapidly. Since pesticides are widely claimed to protect around 40 per cent of the world's food supply,[20] it seems reasonable to infer that global farm pesticide use could potentially double in line with the need to double world food production—and possibly increase even faster than that, as current smallholder organic farms in the developing world are gobbled up by global agribusiness corporations which are wedded to chemicalised production systems. Furthermore, since America currently uses more than one-fifth of the world's pesticides to feed less than 10 per cent of the world's population, it is entirely possible that global farm pesticide use—and hence planet-wide pesticide pollution and food chain contamination—could quadruple from today's levels if US farming systems catch on widely, as they have already done in Latin America. Agriculture accounts for about three-quarters of total world pesticide use, the remaining quarter being used by industry and in the home: all are likely to increase in line with growth in the world economy and consumer demand.

The effects on human and environmental health of this upsurge in pesticide use will coincide and potentially combine with the effects of a doubling (or even tripling) in the volume of most other forms of chemical pollution, such as e-waste, hazardous waste, mineral waste, fossil fuel pollution and so on. The tendency of governments, industry and consumers has been to view these as isolated problems, rather than as components of an aggregate toxic shock to the Earth and to ourselves as individuals. If such attitudes persist, then the scale of this shock will be under-rated rather than over-rated, and under-reported rather than over-reported. Pesticides are already linked in the scientific literature with leukaemia and other cancers, with birth defects, infertility, brain, nerve, skin and developmental disorders, respiratory diseases, diabetes and depression and, at the very least, it seems probable that a doubling in their use will increase the incidence of such effects.

The latest pesticides are depicted by the chemical industry as 'softer', with effects more specific to the target pest, less toxic to humans and gentler to the environment. Whether this is in fact true remains open to scientific debate. A recent study reported that a decline of between 27 and 42 per cent in the number of wildlife species in European and Australian rivers that were contaminated by new-generation pesticides, the neonicotinoids (or 'neonics') compared with uncontaminated streams[21] while a second study warned of their ability to accumulate in soils and poison birds and mammals.[22] The neonics are developed from nicotine and, at a minimum the assumption that they are safer seems questionable in view of widespread scientific acceptance that tobacco causes cancer and other diseases. Both Europe and Australia use the latest pesticides and have strict controls over their use and permitted levels in the environment. Essentially, the studies demonstrated that the controls don't work. Comments ecologist Emma Rosi-Marshall of the Cary Institute of Ecosystem Studies, New York: 'We are at a crisis point, with species loss on a global scale, especially in freshwater ecosystems. Considering pesticides along with other known threats to biodiversity may be crucial for halting species declines.'[23] If the experience of the organochlorines is anything to go by, it may be equally crucial to preventing health and genetic damage in humans, though the evidence on this is not yet in.

The neonics have been implicated in another effect with profound implications for our economic, social and environmental wellbeing: scientists at Purdue University have found they are involved in the mass death of honey bees.[24] Since bees earn the world economy about a third of a trillion dollars every year by pollinating food crops, not only would the economic loss be considerable were their numbers to decline worldwide, but the world food supply would also shrink, causing hunger in some regions and food price hikes for everyone. This illustrates that the 'second round' effects of chemicals can sometimes be more serious, and affect more people, than being directly poisoned. After fierce debate between pro- and anti-pesticide lobbies, Europe imposed a two-year ban on the use of three neonic pesticides[25]—a step that was strongly criticised by industry. The United States took an opposite stance, rejecting calls for a ban and arguing that there were many causes for bee colony collapse.[26] The tussle on both continents highlights the enormous difficulty in identifying and then proving ill-effects from particular pesticide use to the satisfaction of regulators.[27] A central figure in the row, Professor David Goulson of Stirling University, commented:

> There are obvious parallels with the tobacco industry. For 50 years they insisted that smoking wasn't harmful to human health, even when the scientific evidence piled up, they still claimed there was no link. And they funded scientists to come up with spurious studies,

> which seemed to back them up. Here, the scientists funded by the [agrochemical] industry are the only ones that stand up and say, as far as we can tell, there's no effect of these pesticides on bee health. I can't help but being [*sic*] highly cynical about the independence of any of that [research].[28]

Goulson's argument raises an unsettling question: if every new chemical were to undergo half a century of prevarication by manufacturers before its safety or harmfulness can be determined by society, it will be tens of thousands of years before all the dangerous ones are investigated and eliminated. And what will be the human cost of such a delay in the meantime?

Chemical intensification

As a growing number of scientific studies are showing, today's children first encounter petrochemicals in the womb, and then in their mother's milk. Seemingly unaware, parents then clothe their children in garments made from petrochemicals, feed them on food containing petrochemicals, give them toys made from petrochemicals to play with, bathe them in petrochemicals, put them to bed in them and surround them in their daily lives with homes furnished and decorated and cars made of these substances. Then they wonder why their children are becoming increasingly sickly and suffering more unexplained conditions. So they take them to the doctor, who prescribes medication which is, quite often, made from petrochemicals.

A piviotal issue for the future is 'chemical intensification', which is the widespread replacement of natural materials with petrochemicals in industrial and commercial products such as lubricants, coatings, adhesives, inks, dyes, creams, gels, soaps, detergents, fragrances, textiles and furnishings. Indeed many objects in daily life which used to be made of wood, natural fibres, vegetable oils, clay, metal or glass are now made from plastics or synthetic fibres: furniture, tableware, toys, ornaments, clothing and footwear, bedding, shopping bags, packaging, tools, building materials, vehicle and aircraft parts. In the space of a generation society has gone from being largely natural product-based to largely chemicalised. Even 'wood' nowadays consists, more often than not, of wood fibres artificially bound together with epoxy resins: so-called particle board.

'Chemical intensification is not just a measure of the chemical production and use but reflects changes in functions of chemicals and the importance of chemicals in all aspects of economic development. It also incorporates the increased complexity of chemicals themselves and the ever lengthening and more intricate chemical supply chain,' the UNEP explains. Evidently, UNEP foresees substantial scope for negative effects from this ever-increasing presence and volume of chemicals in our daily lives—although it clings to a faint hope for 'sound management' of this rapidly expanding and diversifying chemical chain.[29]

This abundance in chemical diversity and use is likely to propel the expansion of worldwide production at a rate of about 3 per cent a year between now and 2050, UNEP anticipates. This may not sound much, but it is likely to be a faster growth rate than even the world economy as a whole—and implies a doubling and possibly a tripling in worldwide petrochemical use and release into the environment by 2050.

Megakillers

Despite progress in destroying the world's main existing stocks of chemical weapons under the Convention on Chemical Weapons, an estimated 15,000 tonnes of nerve agents and some 4.7 million chemical bombs and shells remain in the hands of Russia, America and two other countries as of 2013: enough to kill every human three or four times over. Six countries suspected of holding chemical weapons—Israel, Myanmar, Angola, Egypt, North Korea, South Sudan—have either not signed or not ratified the Convention, or have not declared their stocks.[30] A recent exception was Syria, which agreed to disclose its stocks to the UN as a first step towards destruction, following heavy international criticism resulting from the use of nerve agents in the country's civil war.

Of growing concern, given what has emerged about the Earth-wide continual dissemination of man-made chemicals, is the fate of tens of thousands of tonnes of chemical munitions dumped into the world's oceans between the end of World War I and 1972, when a convention banning ocean dumping was declared. Nobody knows the exact quantities, locations or types of the dumped chemical agents: these facts are lost to history and thus render any serious clean-up effort impossible.

The US Army alone is said to have 'dumped 64 million pounds [29 million kilos] of nerve and mustard agents into the sea, along with 400,000 chemical-filled bombs, land mines and rockets and more than 500 tonnes of radioactive waste' at twenty-six sites that ring the US coastline.[31] A report to the Congress in 2007 stated that there were thirty-two American dump sites off the US coast, and forty-two off the coasts of other countries, notably

Japan.[32] Around five hundred people, mostly fishermen, have been injured by ocean-dumped chemical munitions in the last half century,[33] but the real concern is over what happens when these ultra-toxins begin to leak out, into the ocean and into the global food chain. Most nerve agents tend to dissolve in water, but substances such as sulphur mustard can remain dangerous for years. The Congressional report concluded: 'Thus far, there have been no comprehensive scientific studies of potential risks to human health and the marine environment . . . Therefore, it is difficult to provide definitive answers to questions about risks raised by public health and environmental advocates, marine conservationists, and the general public.' It should be added that concerns about the impact of ocean-dumped chemical weapons largely predate modern scientific understanding of the way and extent to which man-made chemicals now circulate and disseminate throughout the Earth system via the six pathways (outlined in Chapter 2)—and some of the most murderous substances in human history remain forgotten, but not gone.

Criminal chemicals

The production of chemicals is so profitable that it has become a specialty of organised crime around the world. Illegal recreational drugs are one of the most rapidly growing and uncontrolled sources of toxicity on the planet. While their users no doubt regard them as fairly harmless—or at worst, risky only to themselves—there is in fact a swelling tide of toxic contamination of soil, groundwater and urban waste water systems and even the food chain, with potential to harm millions of people who don't take drugs. This applies especially to amphetamine-family drugs, synthetic cannabinoids and their chemical precursors, which are usually cooked up by backyard operators who know little about chemistry and even less about safe production and waste disposal. The market for these drugs is thought to have grown enormously in recent years due chiefly to rising incomes in Asia, as suggested by a 73 per cent increase in methamphetamine seizures by authorities globally in a single year (2010).[34] Demand for synthetic drugs is even thought to be overtaking demand for traditional 'natural' substances such as opium and cocaine, which suggests that global contamination caused by synthetic drugs is likely to worsen.

The main routes for toxic substances from illegal drug use into the wider environment are in the urine of tens of millions of users and via the improper disposal of chemicals during manufacture, down drains and toilets. Homes and trailers used for 'cooking' these illegal drugs are afterwards permanently contaminated with carcinogens and other toxins; these sites are deemed uninhabitable by local authorities and are now often destroyed.

Recreational drugs are popular because a very tiny quantity can produce quite strong and lasting effects on the neurological system and mind of the user—and it is this very aspect which makes them such a dangerous part of the planet-wide chemical mix. Many of these drugs are endocrine disruptors, meaning that tiny amounts can also interfere with the body's hormonal and reproductive systems, and there is growing evidence that they may be carcinogenic: a US study has found a 'significantly increased risk' for regular drug users of contracting non-Hodgkin's lymphoma.[35] Recreational drugs are also strongly linked to acute effects, such as heart attacks and poisoning, and are also associated with chronic effects such as lung disease and psychosis. A British team has ranked twenty common illegal drugs by degree of harmfulness[36]—but the focus is primarily on users, and their contribution to the 'chemical soup' affecting the general population and environment has so far been inadequately investigated.

Use of recreational drugs generally starts with young people out for a thrill. At this early point, some individuals are open to educational messages about the risks of self-harm. However a necessary additional message, as yet unheard, is that by taking these drugs they are helping to poison society, wildlife and the local and global environment. People who poison water supplies are prosecuted—but for drug users, who also poison water and the environment directly and indirectly, there is no such penalty. There is no easy solution to the global illicit drug problem, but awareness that drug use is now an important contributor to the poisoning of our planet—and thus a practice that is hostile to the future of all humans and all life—may help to deter some young experimenters.

Seeing the unseen

The categories of contamination described in this chapter share one thing in common: they are little known as pollutants by the public at large, and their potential to injure our health and wellbeing into the future is poorly understood. This underlines the urgency for educating a new breed of consumer: not one who witlessly seizes the first enticing product in the supermarket or electronics store, but one who inquires not only into the product and what it contains but also into its past and its future: the processes by which it was made and how it may be disposed of. Without such consumer awareness, humanity is fated to poison itself in ever-increasing doses. Without such consumers, ethical industries will not have the support and economic signals they need to introduce clean, sustainable production methods and products such as those described in Chapter 8. Without such consumers, unethical and 'quick and dirty' industries will put the clean guys out of business. The dirty guys will win—and everyone, including us, them, our children and theirs, will lose.

In Chapter 9 we will explore ways such a process of enlightenment might come about, but for the time being it is enough to be mindful that our future is in our own hands. What that means for the health of humanity is explained in the next chapter.

CHAPTER 6
SICK SOCIETY

The danger we face is not simply death and disease . . .
these synthetic chemicals may be changing who we become.
They may be altering our destinies.

Colborn, Dumanowski and Myers, *Our Stolen Future*, 1996

In the first ten years of the twenty-first century, more than 3000 children were poisoned in the Australian town of Port Pirie, with the full knowledge of governments, health officials, regulators, industry and the community itself.[1] The poisoning had actually been going on for more than 120 years, ever since the first lead smelter was built in the small, regional South Australian industrial centre; the first warning had been sounded in a government inquiry as far back as 1925, with reiterations being made on numerous occasions, particularly in the 1980s and '90s.

'The adverse impacts associated with (lead) production have been consistently downplayed by industry, governments, councils, health officials and regulators,' says Professor Mark Taylor, an environmental scientist at Macquarie University who has investigated the case. 'Even some academics argue the effects of low lead exposures are not of significant concern. Due to ignorance, misinformation, and deliberate obfuscation of evidence, generations of families living next to lead-mining, smelting, and refining centres such as those in Broken Hill, Port Pirie and Mount Isa, have been and continue to be exposed to environmental lead, a known neuro-toxic contaminant.'

His statement was based on a series of ground-breaking studies by scientists over several decades. In just one of these, a 1992 investigation involving 492 children living around the Port Pirie smelter, researchers established a clear correlation between raised lead levels in the children's blood and reduced IQ: children with up to 30 micrograms of lead per decilitre of blood were found to have lost between 4 and 5 per cent of their intelligence.[2] The lead dust lay like a toxic grey mantle across the homes and suburbs surrounding and downwind of the smelter. Samples taken two decades later, in 2011, by Taylor and his team from the children's playground equipment returned average lead values 173 times higher than those at a comparable playground in another nearby town, Port Augusta, which had no smelter. Not content with sampling only the equipment, the researchers took samples from children's hands after they had been playing outdoors for a period of time: 'Hand lead is of greater concern because young children tend to put their hands in their mouths, and it is therefore a significant pathway for childhood lead exposure,' Taylor explains. After just twenty minutes' use of the playground equipment, children typically showed an average daily dose more than forty times above recognised safety limits. Although born with relatively low lead levels in their blood, by two years of age the children were already showing lead contamination above internationally recognised levels of concern—mostly inhaled or ingested in their own homes. The exposure continued as they grew older and began to attend pre-school and junior school.

Taylor explains:

> Childhood exposure to lead has been linked to lower IQ and academic achievement, and to a range of socio-behavioural problems such as attention deficit hyperactivity disorder (ADHD), learning difficulties, oppositional/conduct disorders, and delinquency. The disabling mental health issues from lead exposure often persist into adolescence and adulthood . . .
>
> There has been and continues to be significant government knowledge of the true nature and extent of the problem. The fact that the politicians have not acted on the evidence demonstrates they have ignored the information coming from their staff, and/or have not had the willpower or commitment over the last 30 years to take effective action to eliminate preventable and damaging exposures once and for all.

The authorities responded to the threat to child health and safety by advising the parents to wash the children's hands more thoroughly and to clean their homes more thoroughly—a recommendation made primarily because the smelter was continuing to belch 44,000 tonnes of neurotoxic emissions over the suburbs every year. 'The only conclusion one can draw from the failure to eliminate preventable lead exposure in Port Pirie is that there has been an absence of decisive and competent leadership from successive governments over the last 30 years,' Taylor mournfully observes.

The poisoned children of Port Pirie are far from alone: worldwide, tens of millions of children are being exposed on a daily basis to nerve poisons emitted by myriad sources—lead smelters, coal-fired power stations, toxic garbage heaps, traffic, industrial installations and contaminated food—and are suffering lifelong impairment to their precious, developing intelligences. In the case of common toxins such as lead and mercury, governments are well aware this contamination is happening—or at least have strong grounds for suspecting it—but prefer not to act because of possible detriment to industries that would face the higher costs involved in suppressing emissions, and for fear of political lobbying based, baldly put, on the argument that to stop poisoning children would be harmful to business profits and employment.

A dumber race?

A possible, though unpalatable, explanation for why today's society is so complacent about its increasing chemical immersion—or indeed other major existential threats such as climate change—may simply be that we are less intelligent than previous generations of humans.

Though we pride ourselves in our remarkable technological and creative prowess, our educational standards and our urban sophistication, it is also quite possible that individual IQs are now a few points lower than they were in our great-grandparents' time, as a result of a century of exposure to lead, mercury, organo-chlorine pesticides and other substances known to inflict permanent neurological damage on the brains of the very young, reducing their intelligence lifelong. Although our egos might howl with indignation at the mere idea that, on average, we're not quite as smart as our ancestors, and although the actual loss of intelligence per person is probably small, such a possibility nevertheless highlights the potentially devastating intergenerational impact of exposure to a toxic soup. More importantly, it foreshadows long-term consequences for a civilisation that may be rendering itself progressively less and less intelligent, to the point where it is no longer smart enough to devise and take effective action in the face of numerous threats to its wellbeing, health, social stability and prospects for long-term survival.

Though not widely studied or tested internationally as yet, this scenario of the 'dumber society' is not without an evidentiary foundation. We have known for over forty years that exposure to lead, mercury and organochlorines damages the central nervous system of infants and small children, and reduces their IQ.[3] This is supported by the findings of scientific studies, such as the Port Pirie and Minamata investigations, all around the world.

A loss of two to five IQ points doesn't sound a lot, but even such small reductions in intelligence have been quite strongly linked in the scientific literature to increases in murder rates, violent crimes, juvenile delinquency, reduced school performance and unwanted pregnancies in the most-affected populations. According to Harvard University's Professor David Bellinger, lead alone has caused the loss of twenty-three million IQ points from the brains of American children, organochlorine pesticides another seventeen million and mercury around 300,000.[4] Other scientists conclude that chemical poisoning from these three sources alone is causing more damage to the intelligence of society than premature births, autism, ADHD or traumatic brain injuries, and deserves to be rated alongside other major diseases that afflict modern society.[5] These studies also highlight one of the more insidious aspects of chemotoxicity: while it may not kill the person most affected, it may be an influential factor in them killing, raping or damaging someone else. The risks of such subtle yet life-threatening effects are rarely considered—and impossible to scientifically assess—when new chemicals are released into the Earth system. However, they cause casualties as surely as do other directly toxic substances—the difference is just that you can't sue the manufacturer.

These three substances—mercury, lead and organochlorine pesticides—are here singled out because they have the best-documented case histories for causing developmental disabilities and lifelong mental deficiency. They also especially affect the brains of infants and children while causing little harm to adults. However, as Harvard's Professor Philippe Grandjean has pointed out: 'Two hundred additional chemicals are known to cause clinical neurotoxicity in adults. Despite a lack of systematic testing, many more are known to be neurotoxic in laboratory models. Their toxicity to the developing human brain is not known and they are not regulated to protect children.'[6] In all likelihood the number of chemicals which damage the human brain, mind and intellect runs into the thousands, but too little is known about how they do it for governments to control their production and use. The gaps in our knowledge of the harm caused by these substances, the way they interact with one another plus the high levels of proof required to satisfy regulatory bodies that they should be banned, together make the task of protecting our children and their precious intelligence daunting, if not impossible for the time being.

Furthermore, it is not only 'chemicals' that can harm intelligence—but also the modern industrial food system: a study by Kate Northstone of Bristol University and colleagues found: 'There is evidence that a poor diet associated with high fat, sugar and processed food content in early childhood may be associated with small reductions in IQ in later childhood, while a healthy diet, associated with high intakes of nutrient rich foods described at about the time of IQ assessment may be associated with small increases in IQ.'[7] In their study, children nourished on healthy foods were almost three IQ points smarter than those fed mainly on processed foods.

In a hard-hitting review article in the British medical journal *The Lancet* published in 2014, Grandjean and colleague Philip Landrigan, called on all nations to 'transform their chemical risk-assessment procedures in order to protect children from everyday toxins that may be causing a global "silent epidemic" of brain development disorders'.[8] 'Neurodevelopmental disabilities, including autism, attention-deficit hyperactivity disorder, dyslexia, and other cognitive impairments, affect millions of children worldwide, and some diagnoses seem to be increasing in frequency. Industrial chemicals that injure the developing brain are among the known causes for this rise in prevalence,' they argued.[9] 'Strong evidence exists that industrial chemicals widely disseminated in the environment are important contributors to what we have called the global, silent pandemic of neurodevelopmental toxicity. The developing human brain is uniquely vulnerable to toxic chemical exposures, and major windows of developmental vulnerability occur in utero and during infancy and early childhood.'

They warned that the pandemic was causing reduced national economic growth, reduced earnings, increased health costs, crime, violence and drug abuse, noting that, in the US, murder rates actually fell after lead was banned for use in petrol.

'Current chemical regulations are woefully inadequate to safeguard children whose developing brains are uniquely vulnerable to toxic chemicals in the environment,' the scientists said. 'Until a legal requirement is introduced for manufacturers to prove that all existing industrial chemicals and all new chemicals are non-toxic before they enter the marketplace . . . we are facing a pandemic of neurodevelopmental toxicity.' However, foreshadowing the difficulty which such well-intentioned recommendations are likely to face, the pair found themselves under attack from fellow scientists who criticised their findings as 'not new', 'confused' and 'exceptionally costly'.[10] These responses clearly indicate the opposition which remains to be overcome, in science as well as industry and government, before children can be made safe.

In a society that professes to love its children and aspires to give them the best possible start in life, the harm now being inflicted on their developing minds is out of all proportion compared with only a few decades ago. This raises profound questions: why do we strive to develop excellent education systems, when at the same time we damage the minds that are to receive them? How can we claim to care about our children when, year after year, we increase both the number and volume of the very things that can cause them lifelong harm? Are rising crime, antisocial behaviour and even terrorism rates attributable—even in part—to our failure to protect young minds from chemical damage? And especially, how long can we pretend to ourselves that we are not complicit in this damage, and avoid the truth that the substances which inflict it are the result of our own unbridled demand for consumer goods and an ever-higher material 'living standard', and the unregulated efforts of industry and the market to satisfy us?

The implications of all this are disturbing. The loss of two or three IQ points per person may not seem to make a big difference in the overall scheme of things. But the loss of two or three points *in each successive generation* is a horrifying prospect. It carries the possibility that global society in the mid-twenty-first century could have a 10 per cent lower IQ than, say, the World War II generation. This would have, among other things, a catastrophic impact on the world economy, national economies and industry in terms of lowered productivity: recent economic research finds strong links between intelligence and national output.[11] And, since low IQ is also quite strongly correlated to criminality—the typical criminal is said to have an IQ of ninety-two (eight points below the norm)[12]—a society-wide drop of around ten points in IQ risks an upsurge in murder, violence, imprisonment rates, mental hospital intakes, gang warfare, illicit drugs and international crime as well as an increase in welfare outlays. That such an outcome might stem from increasing chemical damage from all sources combined to the brains of today's children is not a possibility that has been widely canvassed: our society has, for the most part, yet to join the dots.

Terrible waste

The rubbish discarded by modern society comes at a high price. According to a study carried out in India, Indonesia and the Philippines, we need to start thinking of waste in the same way we think about malaria or other big killers: the toll is just as high, and maybe even higher. Investigation of waste dumps led scientists to conclude that in these three countries alone, close to nine million people were exposed to damaging toxins, and were, as a result of disability or premature death, losing a total of between 800,000 and 1,500,000 years of useful life (known as DALYs, or disability-adjusted life years). This contrasts with 725,000 years lost to malarial infections.[13]

'This is the first estimate of the burden of disease resulting from living near toxic waste sites,' said lead author Kevin Chatham-Stephens of Icahn School of Medicine, New York. 'Lead and hexavalent chromium proved to be the most toxic chemicals and caused the majority of disease, disability and mortality among the individuals living near the sites. We were surprised that health impacts of living near toxic sites were on par with other well-known threats to public health such as malaria.'[14]

The team investigated levels of lead in the blood of children living near the dumps and concluded it had cost each child, on average, six to eight IQ points, and increased the number of mentally retarded children by around 6 per cent. In these countries, the money simply does not exist to clean up most of these contaminated sites, thus it is likely that the substances causing this grievous damage will remain mobile and gradually circulate—in water, air, soil, food and goods of trade—around the planet. This emphasises that while the worst impact of hazardous waste may be seen today in developing or newly industrialised countries, the long-term consequences can affect everyone. If we want to live clean, healthy lives, we need to pay attention to what happens in other places around the world—not just our own backyards. And realise that, in a globalised economy, we are the ones who generate the demand that causes the waste that exists in others' backyards.

Childhood calamity

'Children today are sicker than they were a generation ago. From childhood cancers to autism, birth defects and asthma, a wide range of childhood diseases and disorders are on the rise. Our assessment of the latest science leaves little room for doubt: pesticides are one key driver of this sobering trend,' says the Pesticide Action Network's Kristin Schafer.[15] PAN is a worldwide network founded in 1982 and connecting six hundred groups in ninety countries. It explains: 'Pesticides don't respect national borders. Tonnes of agricultural chemicals cross international boundaries every year, either through the international marketplace or carried by wind and water currents.'

Its recent study, *A Generation in Jeopardy*, though based largely on North American data, says 'Compelling evidence now links pesticide exposures with harms to the structure and functioning of the brain and nervous system. Neurotoxic pesticides are clearly implicated as contributors to the rising rates of attention deficit/ hyperactivity disorder, autism, widespread declines in IQ and other measures of cognitive function.'

According to the report:

- Pesticide exposure now contributes to a number of increasingly common health problems for children, including cancer, birth defects and early puberty.
- Evidence connecting pesticides to certain childhood cancers is particularly strong.
- Extremely low levels of pesticide exposure can result in significant health damage, particularly during pregnancy and early childhood.
- Emerging science suggests that pesticides may also be important contributors to the current epidemics of childhood asthma, obesity and diabetes.

Babies are especially vulnerable because, relative to adults, they consume seven times more water, three to four times more food and breathe twice as much air for their body weight, and also have a lower ability to break down and excrete chemical products. This increases their overall risk when exposed to any form of toxin. At the same time their developing central nervous—and hormonal systems are more vulnerable to chemical disruption, with the possibility that early damage can last a lifetime. A trailblazing US study in 2011 by Coleen Boyle and colleagues found that one in six American children now has a developmental disability and that the incidence is increasing, requiring ever-greater health and educational resources (see Figure 4).[16]

The decline in health of the world's children has been drawn to the attention of all governments and international agencies in ways they should be unable to ignore. For example, in 2006

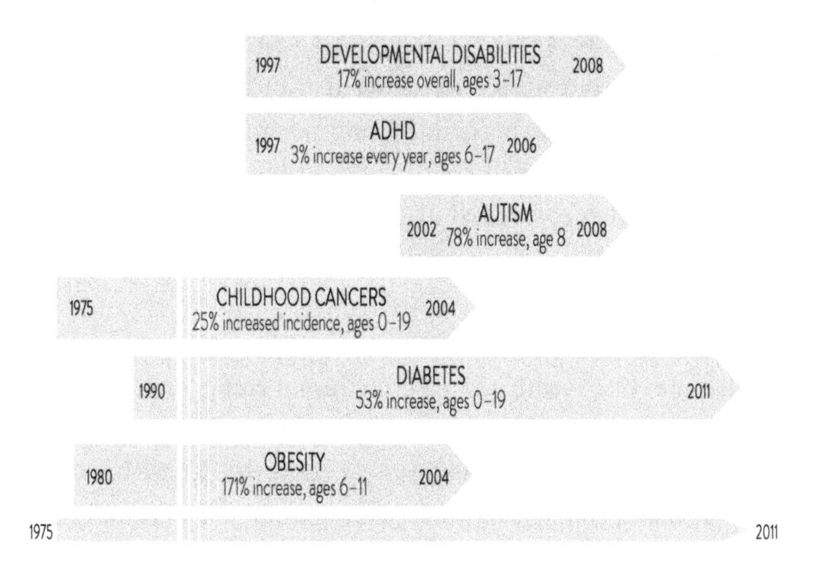

FIGURE 4 Increase in childhood diseases over a generation

Source: Pesticide Action Network (PAN), *A Generation in Jeopardy*, 2012, p. 2.

Professors Philip Landrigan and Joel Forman told a World Health Organization conference: 'Fetuses, infants and children are exquisitely sensitive to environmental exposure [to chemicals]. Recent evidence confirms children are exposed more than adults.' They referred in particular to the work of British Professor David Barker, of Southampton University, who has stated that the early environment is a powerful determinant of a person's health for the whole of his/her lifespan.[17]

'Children are surrounded by a large and increasing number of chemicals. There is increasing evidence of toxic effects and disease,' they said. 'About 80,000 chemicals are produced commercially, and three thousand of these are produced in volumes of more than a million pounds [454,000 kilograms] a year.' No basic toxicity information was available for about half of these high-volume chemicals, the researchers added, and no information on their impact on developing children was available for 80 per cent of them. US studies confirmed that higher levels of chemicals were present in children than in adults, indicating that children were more exposed.

The mass poisoning of children first drew global attention in the late 1950s with the case of the morning-sickness drug thalidomide, which caused birth deformities in an estimated 10,000 to 20,000 children in forty-six countries. There has since been a rolling thunder of published scientific evidence of other chemical-induced effects in the very young, especially involving damage to the developing brain, nervous system, mind and reproductive system.

Landrigan and Forman stated:

> Initial recognition was of obvious, massive damage caused by high-dose exposures . . . But later studies with more sophisticated tools have shown in every case that the brain injury caused by toxic chemicals is not limited to obvious conditions. It is now recognized that there exists an entire spectrum of diminished brain function in persons exposed to toxic chemicals, termed subclinical toxicity.
>
> Widespread subclinical neurotoxicity can affect the health, well-being, intelligence and even the security of entire societies.

Landrigan and Forman pointed out that the loss of five IQ points per child would increase society's number of mentally retarded people by around two-thirds. They add that the impact can be lifelong: 'Early exposure to toxic chemicals may increase risk of degenerative brain disease in later life,' such as Parkinson's disease and Alzheimer's disease. The researchers cautioned that the lion's share of the current ongoing economic cost of childhood diseases, which runs into many tens of billions of dollars, is due to reduced intelligence and lowered productivity: 'Environmentally attributable disease is *very costly to society*' (original italics).

Australian Professor Peter Sly, Deputy Director at the Queensland Children's Medical Research Institute, adds:

> There are increasing data that a lot of the chemicals we use in our homes can have an impact on the developing foetus and child. Unfortunately our environment now is full of chemicals. We have plasticizers, we have flame retardants, we have many other chemicals which remain in the environment for a long time. And mostly we've ignored them. We now are just starting to see some of the chronic childhood diseases may be related to these agents that have been introduced into our environment.

Professor Sly says most new chemical products reach the market without ever being tested for possible child health risks. 'It's not tested at all. There's very little requirement to test a product that comes on the market,' he said. 'There's [*sic*] 40,000 new chemical entities that come onto the market globally every year. The number of those that have actually undergone testing is very few.' He instanced the case of some countries banning BPA in plastic bottles over concerns about its effect on children's health, but added it has now been replaced with other chemicals that have never been tested for their health effects on children at all.[18]

Not paying attention

Attention deficit hyperactivity disorder (ADHD) is today the most common psychiatric or behavioural problem affecting young children in the developed world, affecting between 6 and 7 per cent of children, boys especially. US research indicates that it rose in that country by 3 per cent every year from 1996 to 2006 and by 6 per cent every year from 2003 to 2007. The science to date suggests that something other than the parents' genes is involved in 20 to 40 per cent of cases, and there is a strong relationship between ADHD and a child having organophosphate pesticide traces in their blood,[19] and with exposure to lead. Other possible causes include alcohol intake by the mother during pregnancy, her exposure to tobacco smoke, and chemical food colourings—but these await more evidence. Since a major factor in ADHD appears to be the parents' own genes, then an important unanswered question is: what is causing genetic damage on such a scale across the population, and why is it increasing so rapidly? Clearly, it is something that did not affect many parents half a century or more ago—but which has become more prevalent in recent times.

ADHD cannot simply be dismissed as a case of bad behaviour in youngsters. The World Health Organization lists ADHD as the leading risk factor for acts of murder and other social violence, which claim the lives of quarter of a million young people worldwide every year.[20]

The self pandemic

As every worried parent knows, autism is on the rise—not just in the West, but all around the world, and especially in newly industrialising countries such as China and India. The first cases of autism were reported in 1943 and, while they may not have been the first cases ever, evidently by that time the condition had become common enough to attract medical attention. By 2012 an international study found that worldwide sixty-two children in every 10,000 were affected with autism spectrum disorders.[21] In the US, autism rates rose by 78 per cent between 2002 and 2012, affecting 114 children per 10,000[22] and were thus almost double the global rate. For comparison, rates in Australia were thirty-nine, 181 in Japan and 189 in South Korea. While the recorded increase in the US is probably due in part to greater parental awareness and improved diagnosis, experts in the field consider these reasons explain only about half of the rise: incidence of the condition itself continues to grow rapidly.[23]

The causes of this sudden and dramatic increase in autism are not yet known but, like the suspicions surrounding lead in the 1970s, the finger is starting to point towards various forms of industrial and environmental pollution. For instance, Heather Volk, a researcher at the University of Southern California, and colleagues produced strong evidence of a link between autism rates and living close to a freeway. 'We're not saying traffic pollution causes autism, but it may be a risk factor for it,' was how she guardedly explained her results which essentially showed that if you lived next to a high-traffic area in late pregnancy, your chances of having an autistic child were about double those of parents living in leafier suburbs.[24] Since diesel fumes alone contain at least 650 different chemicals[25] and they in turn are but part of total highway emissions, the task of unravelling which substances may or may not be responsible for conditions such as autism is extraordinarily difficult and may take years. Even then, such a study would not include exposure to a host of other brain-damaging or gene-damaging substances in food, water, furniture and consumer goods.

In 2012 an investigation by the Harvard School of Public Health involving more than 110,000 nurses across the United States concluded: 'Perinatal exposure to higher levels of diesel, lead, manganese, mercury, methylene chloride, and a combined measure of metals were linearly related to increased risk of autism.'[26] Essentially it, too, found that women living in the most highly polluted areas of the country were twice as likely to have an autistic child: it was a second smoking gun.

Scientists have long known that urban air pollution is implicated in much besides autism—cancers, heart disease, lung disease and asthma, cystic fibrosis, genetic and other developmental disorders—and that while it kills around 2.4 million people worldwide every year, conditions are gradually improving in most countries as clean air laws and new engine and industrial technologies catch on. But the case of autism illustrates several important issues: first, that the effects of pollution are often subtle and difficult to attribute to a particular substance or source; second, that effects can be lifelong and widespread in their social and economic costs; and third, that these effects are apparently still increasing *in spite of* the improvements in air quality mentioned above. The third point, especially, implies that while we may have improved the situation regarding certain known toxins, the situation regarding the Earth-wide distribution of toxins, known and unknown, is still deteriorating. Think about it the next time you start a car or ride a bus.

Childhood cancers

Rising rates of childhood cancer are one of the most distressing manifestations of our age: while numbers remain fairly small—175,000 new cases and about 100,000 deaths a year globally—cancers nevertheless have now become the first or second most common cause of child death in developed countries.[27] While rates are highest in the developed world, 'Cancer is emerging as a major cause of childhood death in Asia, Central and South America, Northwest Africa and the Middle East,' says the American Cancer Society. However, the rise in cancer rates has also been accompanied, in countries with advanced medical systems, by major improvements in child survival.

It is frequently claimed: 'The causes of most childhood cancers are unknown'[28]—since the trigger is generally only traced in 5 to 10 per cent of cases. However, Belgian researchers Nicholas van Larebeke and colleagues are convinced that industrial chemicals need to be seriously considered among the possible risk factors:

> Children are at risk of exposure to over 15,000 high-production-volume chemicals and are certainly exposed to many carcinogens. The individual impacts of most of these agents are too small to be detected, but collectively these unrecognized factors are potentially important. Infants and children are exposed to higher levels of some environmental toxicants and may also be more sensitive. During intrauterine development and childhood, cells divide frequently, and the mutant frequency rises rapidly. Endocrine-related cancers or susceptibility to cancer may result from developmental exposures rather than from exposures existing at or near the time of diagnosis.[29]

Childhood cancer rates are generally not increasing quickly—around 0.6 to 1.1 per cent a year in North America and Europe, and somewhat faster in Asia and Africa—but the fact that they are increasing at all suggests that the causes are also becoming more widespread and that no efforts should be spared to pin these down, however difficult or costly it may prove, and eliminate them. In the case of childhood cancer especially, prevention is infinitely more ethical, humane and cost effective than cure. However, far fewer scientific and medical resources are being devoted to prevention rather than discovering 'cures'. The latter are still only partly effective, and one has to wonder why the focus on medical research funding has been on treatment rather than on prevention—and about the influence exerted by large pharmaceutical houses seeking profitable new drugs over contemporary medical practice and policy.

Asthma attack

Asthma rates in children and especially in adults, have soared around the world over the last three decades, and rates in big industrial cities have skyrocketed. An estimated 235 million people now suffer from the condition.[30] For a long time researchers attributed the increase to people living urban lives and not being exposed to immune system priming by pollens and other allergens that would keep them free of the lung disorder; this is known as the 'hygiene hypothesis'. However, gradually recognition has dawned that cities are not particularly clean and healthy—the reverse in fact—and that the rise in asthma has accompanied a general improvement in public health in other respects, so is probably not due to processes or conditions which ordinarily cause the disease, but to something else in the urban environment.[31]

While many national health bodies continue to equivocate about the cause of the increase in breathing difficulties for so many, the World Health Organization is more up front, stating: 'The causes of asthma are not completely understood. However, risk factors for developing asthma include inhaling asthma "triggers", such as allergens, tobacco smoke and chemical irritants.' Its advice: if you want to lower the risk of asthma, then reduce your exposure to air pollution, both indoor and outdoor. If you can.

Mystery plagues

The number of previously unknown and unexplained human diseases is multiplying. Recent examples include Multiple Chemical Sensitivity (MCS), Gulf War Syndrome, Stiff Person Syndrome, Morgellon's disease, Cycle Vomiting Syndrome (CVS), Electromagnetic Hypersensitivity, Chronic Fatigue Syndrome (CFS) and Attention Deficit Hyperactivity Disorder (ADHD). Many of these are still controversial and are generally classified by doctors as MUCs, or medically unexplained conditions—or else are not recognised as diseases at all. For patients with these syndromes, one of the more distressing features of their experience is that doctors frequently attribute the symptoms to a psychiatric disorder, without exploring what in the patient's life and living environment might be leading to their symptoms, whether they are real or imagined. 'Clinicians have developed a range of strategies for deflecting the threat to medical competence posed by medically unexplained symptoms. Generally, these involve shifting the blame from the limits of medicine to some characteristic of the patient. Given our psychologically oriented culture, it is an easy slide from declaring a symptom unexplained to attributing it to specific psychological traits or states of the patient,' Kirkmayer and colleagues explain in a paper on MUCs.[32]

This 'blame the patient' mentality on the part of some doctors may, in part, explain why society has been slow to explore the evidence for chronic chemical poisoning. The tendency has been to encourage patients to blame themselves—either their genes, their imaginations or their mental stability—rather than to mount rational, evidence-based inquiry into other possibilities such as the contribution of the patient's living environment—and the toxins present in it. This is *not* to say that all of these mysterious conditions are linked to chemicals but rather that the possibility is firming that some MUCs may in fact be linked to our immersive daily exposure to thousands of tiny doses of toxic man-made substances—to which our ancestors were *never* exposed. Morality and logic both dictate that we should assess this possibility with the same degree of care, good science and open-minded investigation as for any other possible source of injury.

Morgellon's disease is an intriguing case. According to the Mayo Clinic 'Morgellon's disease is the popular name for an unexplained skin disorder characterized by disfiguring sores and crawling sensations on and under the skin. Morgellon's disease also features fibers or solid materials emerging from these sores.' The explanation continues: 'Fibers found in the sores are usually wisps of cotton thread, probably coming from clothing or bandages. CDC experts note that the signs and symptoms of Morgellon's disease are very similar to those of a mental illness involving false beliefs about infestation by parasites (delusional parasitosis).' Their only clinical advice is 'Get treatment for anxiety, depression or any other condition that affects your thinking'.[33] So, in the Mayo's view, Morgellon's is probably due to a mental state—but if so, what was the trigger? Can we rule out an external factor?

Tens of thousands of Vietnam, Gulf and Afghanistan war veterans would recognise this dismissive response from the medical establishment. Nobody wanted to believe the veterans either, when they argued that their cancers, depression and deformed children might be linked to chemical exposure during their military service. Decades on, and with the persuasive evidence from doctors working with Vietnamese victims that strongly suggests that chemicals can cause health havoc, the prevailing view is changing: while officialdom continues to quibble, both laws and lawsuits are increasingly settling in the veterans' favour, suggesting that society has conceded it wasn't entirely their imaginations.

For civilians the world over, however, such recognition remains in abeyance.

Old plagues ascendant

Over recent years there has been a marked increase in conditions long-recognised by medicine, but previously seen at much lower levels in the population—depression, schizophrenia, learning difficulties, Parkinson's disease, Alzheimer's disease, other mental disorders, auto-immune diseases, asthma, obesity and diabetes—most of which are now to some extent scientifically linked with exposure to man-made chemicals. For example, researchers Anthony Wang and colleagues found California farm workers with high exposure to various common pesticides were three times more likely to suffer Parkinson's disease, particularly the early-onset variety.[34] Similarly a team led by Kathleen M. Hayden of Duke University studied 3000 elderly people in Cache Country, Utah, and found a correlation between pesticide exposure and higher rates of both dementia and Alzheimer's. 'Pesticide exposure may increase the risk of dementia and Alzheimer's disease in late life,' they found.[35] The WHO says there are currently around thirty-six million people living with dementia, and the number is expected to 'more than triple' by 2050.[36] There will be a nation of the demented 100 million strong.

At the same time, chemical toxicity has also been linked by science with the rising incidence of common killers such as heart disease and cancer: a recent US study, for example, links the substance PFOA—commonly found in food packaging, drink bottles and cookware and present in 98 per cent of Americans tested—with elevated cardiovascular risk.[37]

Auto-immune diseases are also on the rise and now affect 5 per cent of the global population, according to the World Health Organization. While the causes of these diseases are complex and they are probably driven by many factors, exposure to man-made chemicals is one of those factors. A WHO report says:

> Both intrinsic factors (e.g. genetics, hormones, age) and environmental factors (e.g. infections, diet, drugs, environmental chemicals) may contribute to the induction, development, and progression of autoimmune diseases . . .

> There is now epidemiological evidence of increasing prevalence of certain autoimmune diseases in highly industrialized countries, which cannot be attributed to better diagnosis alone . . .
>
> There is an urgent need for . . . public health authorities, health professionals, and government agencies to be made better aware of the increasing burden of autoimmune disease due to exposure to physical and chemical agents.[38]

Chemical disruption

In June 2009, the Endocrine Society, comprising 14,000 hormone researchers and other medical specialists in more than a hundred countries, warned that 'even infinitesimally low levels of exposure [to endocrine-disrupting chemicals]—indeed, any level of exposure at all—may cause endocrine or reproductive abnormalities, particularly if exposure occurs during a critical developmental window. Surprisingly, low doses may even exert more potent effects than higher doses.'[39] These endocrine disruptors resemble the hormones which form the body's natural cell signalling system and, by imitating them, send the wrong signals or block hormones' natural access to our cells.

Says the European Environment Agency (EEA):

> Chemicals which disrupt the hormone system . . . may be a contributing factor behind the significant increases in cancers, diabetes and obesity, falling fertility, and an increased number of neurological development problems in both humans and animals . . .
>
> Chemicals which can potentially disrupt the endocrine system can be found in food, pharmaceuticals, pesticides, household

> products and cosmetics. In recent decades, there has been a significant growth in many human diseases and disorders including breast and prostate cancer, male infertility and diabetes. Many scientists think that this growth is connected to the rising levels of exposure to mixtures of some chemicals in widespread use.[40]

'Scientific research gathered over the last few decades shows us that endocrine disruption is a real problem, with serious effects on wildlife, and possibly people,' EEA Executive Director Jacqueline McGlade states. 'It would be prudent to take a precautionary approach to many of these chemicals until their effects are more fully understood.' The Weybridge Study to which she refers showed clear evidence of harm from endocrine disruptors to some wildlife species and in laboratory studies using rodent models for human disorders. 'However, the effects of EDCs on humans may be more difficult to demonstrate, due to the length, cost and methodological difficulties with these types of studies—so wildlife and animal studies may be seen in some cases as an early warning of the dangers.'

The pursuit of scientific understanding is further complicated by the manner in which mixtures of similarly acting EDCs in combination may contribute to an overall effect, whilst exposure to individual chemicals may cause no harm. Also, evidence is emerging that different substances affect people differently, owing to the wide variance in human genetic makeup. Such factors have undermined confidence in the long-accepted dose-response rule, and also make it hard for scientists to identify thresholds of exposure below which there are no effects.

Endocrine disruptors include synthetic oestrogen and well-known toxins such as PCBs, phthalates, parabens, PBDEs, PFOA,

BPA and DDT. These create the body-burden that may accumulate for as long as personal exposure continues—and which will usually decline when exposure ceases. These substances are to be found in such innocuous-seeming items as cosmetics and 'personal care' products, plastic drink bottles, packaging and food containers, household cleaning products, sunscreens, shower curtains, soft toys, TVs, computers and electronic devices, carpets, bedding, wood treatments, medical equipment and medicines, tin cans, detergents and food and drinking water.[41] They are also found in common medical drugs including contraceptives, and in illegal drugs such as methamphetamines. As a result, these chemicals are now disseminated in drinking water and even the food chain, after being flushed down the toilet in the urine of users. In Britain, for example, researchers found traces of the contraceptive hormone EE2 in 80 per cent of the rivers and lakes they tested. The cost of clean-up was estimated at US$40 billion.[42]

The US National Institute of Environmental Health Science makes the point that endocrine-disrupting chemicals are now all around us and that exposure is therefore unavoidable. So sensitive is the human endocrine (hormonal) system that even tiny doses of these chemicals may affect it—levels well below what was once deemed 'no observable effect'. While a handful of EDCs have been banned in a few countries, the overall level of contamination continues to rise sharply throughout the Earth system.

Warnings about EDCs have been growing louder for a whole generation, but are still insufficiently heeded by industry and governments. Dr Theo Colborn, author of the groundbreaking book *Our Stolen Future* and founder of The Endocrine Disruptor Exchange (TEDX) relates:

In 1991, an international group of experts stated, with confidence, that 'Unless the environmental load of synthetic hormone disruptors is abated and controlled, large scale dysfunction at the population level is possible.' They could not perceive that within only ten years, a pandemic of endocrine-driven disorders would begin to emerge and increase rapidly across the northern hemisphere. Today, less than two decades later, hardly a family has not been touched by Attention Deficit Hyperactivity Disorder, autism, intelligence and behavioral problems, diabetes, obesity, childhood, pubertal and adult cancers, abnormal genitalia, infertility, Parkinson's or Alzheimer's Diseases.[43]

Far too little is being done worldwide to prevent endocrine disruption to humans on a large scale, say the UN Environment Programme and World Health Organization, whose joint 2013 report states:

- Human and wildlife health depends on the ability to reproduce and develop normally. This is not possible without a healthy endocrine system.
- Many endocrine-related diseases and disorders are on the rise.
- The speed with which the increases in disease incidence have occurred in recent decades rules out genetic factors as the sole plausible explanation.
- At least 800 chemicals are either known or suspected to be capable of interfering with hormone receptors, hormone synthesis or hormone conversion.
- Numerous laboratory studies support the idea that exposure to chemicals contributes to endocrine disorders in humans and wildlife.

- The risk of disease due to endocrine-disrupting chemicals may be significantly underestimated.
- Worldwide, there has been a failure to adequately address the underlying causes of trends in endocrine diseases and disorders.[44]

The report noted that up to 40 per cent of men in some countries had poor semen quality, and the incidences of genital malformations, premature babies and neurobehavioural problems in children had all risen. 'Global rates of endocrine-related cancers (breast, endometrial, ovarian, prostate, testicular and thyroid) have been increasing over the past 40 to 50 years. The prevalence of obesity and type 2 diabetes has dramatically increased worldwide over the last 40 years,' it added. The EWG has published a list of the main endocrine disruptors along with advice to consumers on how best to avoid them—although, regrettably, this is not always possible.[45]

Cancer growth

Cancer has become the killer of our Age, in both the developing and developed worlds. It costs around US$1.2 trillion a year to diagnose and treat. The WHO's World Cancer Report 2014 reported the global cancer burden is growing at 'an alarming pace' with a 70 per cent increase expected by 2035. 'In 2012, the worldwide burden of cancer rose to an estimated 14 million new cases per year, a figure expected to rise to 22 million annually within the next two decades. Over the same period, cancer deaths are predicted to rise from an estimated 8.2 million annually to 13 million per year.'[46]

> 'Despite exciting advances, this Report shows that we cannot treat our way out of the cancer problem,' states Dr Christopher Wild, Director of the International Agency for Research on Cancer (IARC) and co-editor of the study. 'More commitment to prevention and early detection is desperately needed in order to complement improved treatments and address the alarming rise in cancer burden globally.'[47]

Overall, the rate at which people are developing cancers appears to be increasing at between 1 and 3 per cent per year. In developed countries, one in every three women and one in every two men will be diagnosed with a cancer.[48]

While tobacco smoking, obesity and poor diet (all of which have chemical exposure components) are listed as the main risk factors for contracting cancer, chemical exposure in the home, workplace or living environment is also acknowledged as a significant risk—though one which remains largely unquantified.[49] The US Department of Human Health Services in its *2011 Report on Carcinogens* lists fifty-one chemicals and groups of chemicals which are known to cause cancer and 151 substances or groups which are strongly suspected carcinogens, including many that are commonplace in daily life and consumer products.[50] Just how many cancers are attributable to man-made chemicals, nobody knows. However, doctors estimate that at least half of all cancers are preventable.

Breast cancer is the most common cancer in women, claiming more than half a million lives worldwide every year. Its incidence is rising partly because women are living longer lives—and partly due to other factors, including exposure to 'a cocktail of carcinogens and endocrine disruptors every day that puts us at greater risk' in food, cosmetics, personal care, furnishings and household cleaning products, according to Jeanne Rizzo, CEO of the US-based Breast Cancer Fund. The Fund lists the substances and products most implicated in the disease on its website.[51]

Rizzo was part of the US Congress appointed Interagency Breast Cancer and Environmental Research Coordinating Committee which in 2013 took the important step of calling for a national breast cancer prevention strategy, along with a major increase in research efforts to pinpoint the causes.[52] However, since a genuine and thorough prevention strategy probably entails discontinuing a vast number of household, food and cosmetic products—or at least radically changing their formulation and packaging—it remains to be seen whether US legislators will decide the health of American commerce is more important than the health of American women; this dilemma now faces governments the world over.

Groups such as the Breast Cancer and Chemicals Policy Project at the University of California at Berkeley are developing new ways to assess the risk: 'Chemicals used in industrial processes or found in the environment, consumer products, or workplaces must be tested for their possible impact on breast cancer risk. Testing should identify alterations in biological processes relevant to breast cancer.'[53] One of its leading researchers, Dr Sarah Janssen, argues:

> The burden shouldn't be on the consumer, you shouldn't have to have an advanced chemistry degree when you go to the grocery

> store, or to the drugstore to buy your personal care products or to buy your food to know how to read long chemical names and whether or not they are going to increase your risk of one type of disease or another. The regulations should be much more upstream and the government has the responsibility to ensure that chemicals are safe before they are ever put on the market.[54]

Survival rates from cancer of the breast, bowel and other organs are improving in affluent countries. Ironically this is due to products of the petrochemical and pharmaceutical industries—industries whose other products are increasingly suspected of causing much of the problem in the first place. This raises the unsettling scenario that, if we can eventually cure most cancers chemically, it may render it less important to prevent chemicals from causing cancers in the first place—possibly quelling the impetus to reduce our use of chemicals that cause other debilitating but non-lethal diseases.

Cancer, in other words, is at risk of becoming the acceptable price of living in a chemicalised society.

Depression

The human brain is an organ that is exquisitely sensitive to chemical signals at minute concentrations. The World Health Organization estimates 350 million people now suffer from clinical depression, adding 'The burden of depression and other mental health conditions is on the rise globally.'[55] The University of California Los Angeles suggests about 5 per cent of the US population currently suffers a depressive illness.[56]

Depression has long been described as 'a chemical imbalance in the brain' but this statement really explains nothing, including what causes it. As Harvard Medical School points out:

> Research suggests that depression doesn't spring from simply having too much or too little of certain brain chemicals. Rather, depression has many possible causes, including faulty mood regulation by the brain, genetic vulnerability, stressful life events, medications, and medical problems. It's believed that several of these forces interact to bring on depression. To be sure, chemicals are involved in this process, but it is not a simple matter of one chemical being too low and another too high. Rather, many chemicals are involved, working both inside and outside nerve cells. There are millions, even billions, of chemical reactions that make up the dynamic system that is responsible for your mood, perceptions, and how you experience life.[57]

Cynics have claimed the 'chemical imbalance theory' was created to sell more antidepressant chemicals.[58]

Evidence that man-made chemicals may be implicated in the chain of events that cause the neurotransmitters in our brain to malfunction, resulting in depression, is accumulating—but the scientific groundswell is far from definitive as yet. One suggestive example involved the study of farm workers who were poisoned by organophosphate pesticides, which are known to cause neurological damage: US researchers concluded that 'exposure to pesticides at a high enough concentration to cause self-reported poisoning symptoms was associated with high depressive symptoms independently of other known risk factors for depression among farm residents'.[59] In another case, more than half the residents of 1800 homes in a small Mississippi town who were exposed to pest control chemicals over ten years were diagnosed as depressed years after the event.[60] The fact that certain medical drugs—including some antidepressants!—also appear to produce depression as a side-effect lends weight to the view that depression can be a chemical-induced illness, but there is as yet insufficient research to answer the question 'How much of the depression suffered by today's society is attributable to man-made chemicals?' Answering it is urgent: beside the personal suffering and economic loss it causes, depression is implicated in around one million suicides worldwide each year and twenty million attempted suicides, according to the WHO.

Chemical obesity

'It is a commonly held view that obesity is all to do with too many calories taken in and too few expended in exercise, with a genetic predisposition in some individuals. However, a new line of research suggests that exposure to certain man-made chemicals in our environment can play an important role in the development of obesity. While obesity is a known risk factor for diabetes, evidence is growing that chemical exposures are also implicated in diabetes,' two eminent medical scientists, Spaniard Miquel Porta and South Korean Professor Duk-Hee Lee wrote in a recent review article.[61]

In a review article in the journal *Hormones*, Retha Newbold of the US National Institutes of Health concurred:

> Environmental chemicals with hormone-like activity can disrupt programming of endocrine signaling pathways during development and result in adverse effects, some of which may not be apparent until much later in life. Recent reports link exposure to environmental endocrine disrupting chemicals during development with adverse health consequences, including obesity and diabetes. These particular diseases are quickly becoming significant public health problems and are fast reaching epidemic proportions worldwide.[62]

The World Health Organization says that the global rate of obesity has nearly doubled since 1980. More than 1.5 billion adults are overweight, and of these, 200 million men and 300 million women are obese. More than forty million children under the age of five are overweight. Obesity is the fifth leading risk factor for early death and claims about 2.8 million lives each year. It is no longer mainly concentrated in the affluent countries but is spreading rapidly throughout the world. Diabetes, to which obesity is strongly linked, is also increasing rapidly with 347 million diabetes sufferers and around 3.4 million deaths due to diabetes per year.[63]

Obesity and diabetes are further examples of the tendency of the medical establishment to encourage patients to blame themselves, rather than inquire into more complex and subtle explanations for the drivers of these modern pandemics—and risk offending powerful commercial interests. While the lifestyle and dietary choices of individuals—influenced by the torrent of unhealthy food advertising to which they are daily subjected—have no doubt played their role in the ill-health and death of millions, many scientists are now convinced that other factors are also at work.

Porta and Lee, for example, say in their review that there is plenty of evidence that chemicals found in food and the human environment can cause weight gain in laboratory animals. Suspect substances include POPs, PCBs, pesticides, flame retardants, BPA, phthalates, lead, nicotine, diesel exhaust and medical drugs—all of which are also linked to a host of other conditions. 'It is likely that there are other chemicals in the environment that increase the risk of obesity, which have yet to be recognised,' they add. The biochemical mechanisms involved are not yet clear, but probably include scrambling the body's natural energy storage and distribution system—its hormonal signalling—and damaging its genes. The latter raises the disturbing possibility that conditions such as obesity and gene damage caused by exposure to synthetic chemicals can now be passed to our offspring. Porta and Lee conclude, 'the concern that chemicals in the environment may be partly responsible for the increasing occurrence of obesity in human populations is based on a significant and growing number of mechanistic studies and animal experiments, as well as on some clinical and epidemiological studies. The weight of evidence is compelling.'

If chemicals are partly to blame for obesity, then they are also likely to be partly responsible for the universal upsurge in diabetes, Porta and Lee add. 'Evidence suggesting a relationship between human contamination with environmental chemicals and the risk of type 2 diabetes has existed for over 15 years, with the volume and strength of the evidence becoming particularly persuasive since 2006 . . . Chemicals linked to type 2 diabetes in human studies are POPs (including dioxins, PCBs, and some organochlorine pesticides and brominated flame retardants), arsenic, BPA, organophosphate and carbamate pesticides, and certain phthalates.'

Their overriding conclusion was that, 'given the current epidemics of obesity and diabetes, action to reduce exposures to many chemicals possibly implicated in obesity and, more certainly, in diabetes, is warranted on a precautionary basis'.

Newbold added: 'Public health risks can no longer be based on the assumption that overweight and obesity are just personal choices involving the quantity and kind of foods we eat combined with inactivity. It is quite possible that complex events, including exposure to environmental chemicals during development, may be contributing to the obesity epidemic.'

Preventable problem

In all, the WHO estimates, from one-quarter to one-third of the global disease burden is attributable to 'environmental factors' (as distinct from infectious or genetic causes or accidents), including man-made chemicals and wastes, and that this burden falls especially heavily on children aged less than five years.

UNEP states: 'Exposure to toxic chemicals can cause or contribute to a broad range of health outcomes. These include eye, skin, and respiratory irritation; damage to organs such as the brain, lungs, liver or kidneys; damage to the immune, respiratory, cardiovascular, nervous, reproductive or endocrine systems; and birth defects and chronic diseases, such as cancer, asthma, or diabetes. The vulnerability and effects of exposure are much greater for children, pregnant women and other vulnerable groups.' Neither UNEP nor WHO are organisations prone to exaggeration or to making claims unsupported by solid scientific evidence; their warnings should be taken seriously.

While the evidence for the pandemic of chronic poisoning of humanity is fragmented, the dramatic multiplication of its fragments and their combined weight now point to a clear connection between rising chemical exposure and declining physical and mental health, declining intelligence and premature death in many nations and sectors of the world population. Comments Australian medical practitioner and toxicologist Dr James Siow: 'The cumulative effects of combined toxicants have reached epidemic levels in the 21st century . . . The current persistence of pollutants and their impact on the human genome globally . . . implies that almost every human being on earth can be considered physiologically and biochemically polluted.'[64]

In short, poisoning by man-made chemicals has become the biggest *preventable* healthcare issue of our time—and is due to grow bigger still.

Parental curse

This burden of ill-health is now in the process of transfer from one generation to the next by *epigenetic* means. For many years it was believed by scientists that our genes were fairly stable entities, but researchers are finding increasing evidence that when genes become chemically damaged or silenced (disabled), their behaviour (or 'expression') changes and this can lead to disease. This constitutes a paradigm shift in our understanding of genetics and heritability. Put simply, the current understanding is that the underlying structures of the genes themselves remain constant—but the instructions they send to the rest of the body can be distorted. 'Epigenetic changes can result from exposure to environmental hazards such as cigarette smoke, arsenic, alcohol, phthalates, BPA, as well as other chemicals,' says Toxipedia, a *pro bono* online toxicology encyclopaedia.[65] Milan University researcher Andrea Baccarelli adds metals (such as chrome VI), air pollution, endocrine disruptors and dioxins to the possible causes.[66]

Furthermore, this damage appears to pass from parent to child. 'There are now many examples of epigenetic inheritance through the germ line,' says Australian scientist Robin Holliday, an expert on the ageing process, who has pursued the heritability of epigenetic defects since the mid-1980s.[67] Among the more striking findings is that a mother's diet during pregnancy can affect the risk of her child becoming obese, and that as more and more parents are overweight, the risk of obesity compounds with each succeeding generation.[68] In general, it suggests any chemical damage to her genes sustained by the mother may potentially lead to her unknowingly reprogramming the DNA of her unborn child.

'More than 13 million deaths every year are due to environmental pollutants, and as much as 24% of diseases are estimated to be caused by environmental exposures that can be averted,' is how a team of Chinese, American and Italian researchers sum it up. 'In a screening promoted by the United States Center for Disease Control and Prevention, 148 different environmental chemicals were found in the blood and urine from the US population, indicating the extent of our exposure to environmental chemicals. Growing evidence suggests that environmental pollutants may cause diseases via epigenetic mechanism-regulated gene expression changes.'[69] This research team also published a table listing the various diseases now linked by science to epigenetic effects—a list which includes many cancers (leukaemia, breast, prostate and bowel cancer), schizophrenia, heart disease, traumatic brain damage, neurological disorders, Parkinson's disease, asthma and psoriasis.[70]

Like some biblical curse, chemical poisoning and the harm it causes may thus not only affect the people immediately exposed but also the next generation . . . and the next . . . and the next.

The case for precaution

The evidence presented in this chapter is but a tiny sample of that available in the world scientific and health literature. Together it makes a compelling, if not irrefutable, case for precaution and prevention with regard to the chemicals we release into our environment and put into our bodies. It calls for far greater care over the introduction of new substances with unknown effects, and the reader is encouraged to make their own further inquiry into what the science is telling us. In addition to the desire to alleviate the burden of ill-health and premature death in all societies, we have a duty of care to our children: not to leave them with damaged genes and a fouled world in which they will struggle to lead healthy lives.

What is of particular concern is that it still appears necessary for many hundreds or thousands of people to be damaged, suffer and die before sufficient scientific and regulatory attention is drawn to a chemical event or substance of concern, a proper investigation carried out and the long (and frequently unsuccessful) process of restricting or banning its production and use begins, first in one country and then in others. In the present age, the burden of proof that a chemical is dangerous still rests with its victims—rather than the onus being placed on the producer or commercial user to demonstrate its safety. That governments have displayed a tendency to side with industry against the victims of possible chemical poisoning makes matters even worse—amounting to a governmental decision, whether knowing or unconscious, to tolerate an increasing burden of ill-health and premature death in the community as the acceptable price of 'economic progress'. In any case, this is poor logic—because the rising burden of chemical disease may now be undermining economic growth as well as exacerbating healthcare costs and increasing the taxes needed to pay for them. That the medical profession has in the main behaved as if unaware of the mounting evidence for chronic chemotoxicity in the population, has frequently tended to blame the patient's own genes or mental state and has habitually responded by administering more powerful chemicals is a particular tragedy.

The irony is that this burden of chemical-induced ill health and mental impairment is now becoming so large it may actually be starting to retard economic progress on a global scale. The World Health Organization recognises such an impact may exist, and has created tools to assess the true costs, though these have yet to be widely applied. As a World Bank official remarked: 'The lack of quantification and valuation of environmental health hazards prevent any dialogue on the issue. It's when you put a figure on the environmental health burden of disease that you can talk with decision-makers especially the Ministry of Finance.'[71] In short, until some enlightened economist does the sums, the problem won't exist in the dollar-driven minds of governments and bureaucracies. As UNEP puts it: 'Disastrous incidents make the headlines, but the true costs of chemical mismanagement are dispersed and hidden throughout the population and over time. Such costs are typically carried by a nation's social welfare system and individuals.'[72]

However, proof that we are all physically subject to a complex toxic assault may perhaps soon be quantified using the 'exposome' concept put forward by Britain's Dr Christopher Wild, director of the International Agency for Research on Cancer, in 2006. Essentially, this proposes assessing the combined toxic chemical impact on the entire human genome—our full complement of genes—and hence, for the first time, may represent a way to understand our total burden due to these substances, and their long-term health consequences for the population.[73] Further evidence could also emerge from current large-scale scientific studies to determine what chemicals are found in up to a million babies and children. These studies are now being carried out in seven countries, as described in Chapter 3. These data, when gathered and analysed, will provide a new way for humanity to understand the causes of health conditions which are currently characterised as mysterious and inexplicable, or else are being played down or hushed up by the authorities.

However, understanding is one thing—prevention of chemical harm is another. Harvard's Professor Grandjean notes:

> Prevention . . . will require fundamentally new approaches to control of chemical exposures. The vulnerability of the human nervous system and its special susceptibility during early development

> suggest that protection of the young brain should be a paramount goal of public health protection . . . exposure limits for chemicals should be set at levels that recognise the unique sensitivity of pregnant women and young children, and they should aim at protecting brain development. The precautionary principle . . . would suggest that early indications of a potential for a serious toxic effect, such as developmental neurotoxicity, should lead to relatively severe regulations, which may later be relaxed, should subsequent documentation show that the hazard is less than anticipated.[74]

As we shall see, prevention needs so much more than well-intentioned recommendations and regulations: it needs us.

CHAPTER 7
GETTING AWAY WITH MURDER

The future depends on what you do today.

Mahatma Gandhi

The strange epidemic at Minamata first came to medical attention in early 1956. It was reported, almost immediately, by a doctor employed at the chemical company's hospital, to the management and Board of the Chisso Corporation. He was formally instructed to cease his inquiries into the source of the disease (although, courageously, he did not do so). By 1959 the local university had traced the culprit: organic mercury, one of many contaminants poured by the company into the bay on which the local fishing industry and families depended for their livelihood and food. Angry fishermen stormed the factory offices and were met with blows and threats. Over the ensuing years of legal and physical battles some were bought off with meagre compensation; others continued to fight.

But the pollution did not cease—and nor did the spreading sickness, with all its hideous symptoms and agonising deaths. Over the intervening years, the company denied, lied and flung counter-accusations. It falsely claimed to have solved the problem. It bribed and threatened, set citizen against citizen, families against themselves, unionist against unionist. It hired thugs and lawyers and used them against protesting victims of the poisoning. It invoked the protection of sympathetic local and national government officials. It barricaded its headquarters with steel bars to keep the angry protesters at bay. It met media inquiries with bluster and disinformation. It split the Japanese nation. Eventually it palmed off the entire issue onto a government board of inquiry which returned equivocal findings. Finally, in 1973, Chisso ran up against an incorruptible judge who found the corporation guilty, saying:

> a chemical plant, in discharging the waste water out of the plant incurs an obligation to be highly diligent; to confirm safety through researchers and studies regarding the presence of dangerous substances mixed in the waste water as well as their possible effects upon the animal, the plant, and the human body, always availing itself of the highest skill and knowledge; to provide necessary and maximum preventative measures . . . in the final analysis . . . no plant can be permitted to infringe on and run at the sacrifice of the lives and health of the regional residents.

After a battle that had lasted nearly twenty years, the company president finally apologised to the families his corporation had knowingly maimed and ravaged, with his forehead pressed to the ground in the traditional Japanese gesture of abject submission.[1]

In 1972 Eugene and Aileen Smith, who had followed and reported on the closing chapters of the human drama at Minamata, returned to the US where, almost immediately, they were contacted by concerned citizens in Canada, recounting a tale of identical horror involving a chemical plant that was poisoning a local river, and the indigenous Canadian people who depended on its fish.

And so, on and on the cycle goes.

Time and again, around the world, the pattern has been repeated: a toxic release followed by corporate denial, misdirection, counter-attack, media battles, protracted lawsuits (sometimes conclusive but usually delayed, costly and inconclusive, with much stalling); the complicity and corruption of government departments and officials, community anger, bitterness and suffering. Events surrounding the nuclear disaster at Fukushima, Japan,[2] and revelations post-*Erin Brockovich* around the presence of chromium VI in the drinking water supplies of more than thirty US cities[3] show that the ghost of Minamata has not been laid to rest, and continues to haunt the world.

Bureaucratic intervention

The affair at Seveso, Italy, in 1976 involving a company, ICMESA, a subsidiary of Swiss industrial giant Hoffman LaRoche, exposed an ugly aspect to such events—and the role of governments in trying to silence community concern. A day after the accidental escape of a dangerous gas during the production of herbicides the company initially acknowledged the release had taken place. On day three, local health authorities announced there was 'no fear of any danger to the people living in areas surrounding the plant'. On day seven, the product (TCDD) was revealed to contain the carcinogen dioxin. On day twelve, the local government demurely stated: 'At this time there is no cloud of toxic gas,' and the following day said: 'Other health measures should not be considered necessary or urgent.' Meanwhile, the Regional Health Director proclaimed: 'Everything is under control'—however, *on the same day* the company's own medical director stated, 'the situation is very serious and drastic measures are called for'. Caught in a lie, the local government initially tried to throw doubt on the medical director's credentials, then backflipped on the following day, announcing that 179 people should be evacuated.[4] In all, six tonnes of deadly TCDD were dropped on an area of 18 square kilometres, exposing 37,000 people. Subsequent investigations revealed about 200 cases of serious chemical-induced skin inflammation and an unexplained increase in lung diseases, diabetes and certain cancers, notably breast cancer among younger women. Five employees—including two managers—of the chemical plant were charged, convicted, and then released on appeal without penalty. Twenty years later, in 1996, the event prompted the European Union to pass a law known as the 'Seveso Directive' requiring higher standards from industry for the protection of the public.[5]

Evading responsibility

In India in the 1960s hunger was the focal issue. Ten of millions of people were starving, and its rulers saw a desperate need to build the world's largest democracy into a modern industrial state in order to create jobs for displaced people who were flooding into the cities from rural areas. In the regional city of Bhopal, in what was then Madhra Pradesh State, a solution to both was offered by Union Carbide (India) Limited (UCIL), an offshoot of the US Union Carbide company, which in the late 1960s began construction of a major chemical plant for the production of the pesticides that would be used to protect India's crops from attack by insects, thereby helping to relieve the food crisis. In the poorly planned city whose population had doubled in size in less than twenty years, heavy industry jostled side-by-side with residential and commercial areas. Thousands of land-hungry squatters erected their homes within metres of the chemical plant perimeter.

In 1969 the plant was licensed to produce 5000 tonnes a year of a carbaryl-based insecticide, but a string of technical failures hampered development of the insecticide production process over the ensuing decade. By the time the precursor chemical production unit was fully operational, demand from Indian farmers had moved in favour of cheaper, imported pesticides: production became uneconomic and was scaled back. The more money the plant lost, the more the quality of its operations deteriorated; skilled workers left and were replaced by inexperienced staff. In the crucial methyl isocyanate unit, the workforce was cut from the recommended level of three supervisors and twelve workers per shift to one supervisor and six staff. In December 1981, there was an ominous warning: a gas leak that led to fatalities.

In late 1984, the situation was compounded when India's then Prime Minister, Indira Gandhi, was assassinated and the news caused riots to break out across the city—making it even harder for employees of the plant to get to work. Meantime, a string of safety precautions began to break down around the two tanks that held the remaining sixty-two tonnes of methyl isocyanate. Somehow, water from a washing operation got into pipes linked to the chemical storage tanks and backed up, causing pressure in the tanks to build dangerously over several hours. Soon after midnight it reached a critical level—and erupted. Safety systems designed to trap or neutralise the agent failed and roughly forty tonnes of highly poisonous gas burst into the surrounding atmosphere. Alarms sounded. Workers rushed about attempting to fix things. Sleeping families in the surrounding suburbs awoke to the smell of gas, panicked and began to flee in all directions. Emergency services, broadly, failed to respond, but an Indian Army engineering unit—summoned by the manager of a neighbouring chemical plant—kept its head and began to organise an evacuation of the plant and the transfer of poisoned people to surrounding hospitals.

'Though there was defoliation of trees and some additional contamination of soil and lakes, the main impact of the accident was death and injury to humans and animals. Estimates of the number of immediate human deaths caused by the Bhopal gas cloud vary from the official Indian government figure of approximately 2000 to the 10,000 favored by local activists,' wrote Massachusetts University's M.J. Peterson in his authoritative analysis of the disaster. Another 200–300,000 were injured.[6] Subsequent litigation established 3818 fatalities as due directly to the gas leak, but thousands more people were left crippled for life and many widows and orphans suffered great hardship. The brawl over compensation, justice and responsibility drags on to this day: activists claim the death toll has risen to 25,000 with a further 120,000 suffering chronic ill-health.

The Indian government, which had sought compensation of $3.3 billion on behalf of the victims, settled for $470 million, which was finally paid over by 2003, much to the discontent of many victims. Bhopal was the world's worst industrial disaster, far exceeding in scale a Chinese coal mining disaster that claimed the lives of 1500 miners at the Benxihu colliery in 1942. Fearing retribution, Union Carbide chairman Warren Anderson and the senior Indian government official fled the scene and were never brought to justice: the campaign to do so continues.[7]

Some lessons

Bhopal, Seveso, Minamata and other famous cases have established a pattern for chemical poisoning disasters. It usually falls to the victims of poisoning to prove they have been injured by a chemical event and to demonstrate a clear technical link between cause and adverse health effects, pitted against the frequently combined efforts of corporation and government to hinder and discredit them. Rarely is the suspect factory or industry required to demonstrate that its processes and products are safe, or that it upheld high standards of public safety at the time of the event. Rarely are its management or owners brought to justice. Furthermore, while compensation and apologies may ultimately be extracted after years or decades of duress, they seldom make good the lives laid waste.

The significance of these classic 'chemical battles' between large and powerful corporations and small groups of affected citizens is that they have established an unproductive pattern for conflict and confrontation, in which a company typically digs in, denies responsibility and fights through the courts—leading to surges in anger and hostility from the public, environmentalists and the media. As a result, the prospects for resolving chemical contamination issues rationally and without confrontation are much diminished. Fearful of public outrage or monetary loss, industry seals itself behind lawyers, spin doctors and razor wire, while governments frequently seek to muddy the waters for their own protection, rather than seeking to resolve the problem. Meanwhile, an outraged public and angry green groups frequently deny the industry a reasonable chance to explain, work through the issues constructively and propose acceptable solutions. In a climate of wrath, blame and confrontation, satisfactory solutions are bound to be hard to come by—and thus the chemicals that cause harm are likely to remain in unsafe use and circulation. Fearful of such conflicts, even well-run companies and industries often take refuge in secrecy or flee offshore, making the task of preventing future chemical leaks and associated casualties doubly difficult.

A second feature of significance in these classic 'chemical confrontations' is that they are generally local in character, and provide no model for dealing with chemicals that have disseminated worldwide in the Earth system, or chemicals which have combined with other substances to deliver secondary toxic effects. Currently, companies may be prosecuted for polluting the local river—but not for polluting the planet or human species. People in one country seldom have any legal remedy or recourse against chemical fallout from another country or continent, especially if it is emitted by a range of industries or by pervasive activities such as burning coal to make electricity, concrete or steel.

While there are valuable lessons to be gleaned from every accident, and while better regulations within particular jurisdictions often result after an accident, such is the rapidity of chemicalisation in every aspect of human life that the hope of preventing all similar disasters in future is small—and the prospects of fully containing a growing global toll attributable to population-wide chronic chemical exposure, very low indeed. Above all, these stories teach us that dealing with the global chemical flood on a case-by-case basis will not resolve the problem. More effective, visionary, far-reaching and cooperative solutions are called for.

Dirty dozen

Of the many tens of thousands of toxic and carcinogenic chemicals which have been developed and released globally in the past century of industrial development, only a tiny handful have ever been banned. The Stockholm Convention on Persistent Organic Pollutants was established in 2001 by the United Nations to eliminate chemicals that persist in the food chain and environment: the Convention has 179 parties, including the European Union, which means that most countries in the world support it.

In 2004, the Convention outlawed nine of a set of twelve chemicals dubbed the 'dirty dozen', and called for the remaining three to be restricted. It subsequently added a further nine substances in 2010, adding up to a grand total of twenty-one substances banned or restricted over a period of twelve years—almost half a human generation.[8] Even so, the bans only apply in countries that signed and ratified the convention.

At such a rate of progress it will take more than 50,000 years to assess and eliminate all currently known or suspected man-made toxins. This is equal to the entire span of existence of *Homo sapiens sapiens*. Even then such substances will only be banned in certain countries, which in turn will be virtually powerless to prevent pollutants' entry via the six global pathways and the resulting exposure of their citizens. This estimation of progress also takes no account of all the tens of thousands of new chemicals to be released in the meantime.

The picture is bleaker at national level. In America, for example, it is claimed that on only five occasions in thirty-six years has the US EPA succeeded in having a new chemical banned.[9] Since America admits to using more than 84,000 chemicals—about one-third of which are considered potentially toxic, carcinogenic or gene-damaging—and since a further one thousand or so new substances are being created and released each year, it would appear that most of today's newly developed chemicals are routinely approved for release regardless of their potential toxicity or ultimate human and environmental cost, even in well-regulated societies. Regrettably, the widely used phrase 'dirty dozen' appears to have created an impression in the minds of many citizens that the number of toxic chemicals is very small, and hence fairly easily reined in: however in reality there are in excess of 30,000 known or suspected toxic substances totalling ten million tonnes in volume (according to the UNEP's 2012 data), plus a host of unintended pollutants created by mining, manufacturing, energy production and other activities.

Weaker controls

While global chemical use is forecast to intensify, growing by around 3 per cent per year up to 2050, the world's ability to regulate and limit use is likely to weaken.[10] In the first decade of this century, chemical output in Asian countries grew three to five times faster than in North America and Europe; this trend has continued through the 2010s. Output in well-regulated North America and Europe, for example, is forecast to grow by 25 per cent up to 2020. However, chemical output in Asia is predicted to increase by 46 per cent, China's output by 66 per cent and India's by 59 per cent, according to the UNEP.

In the beginning, big chemical companies saw the developing world as a promising new market for chemical products which were difficult to sell, or in some cases were banned, in developed countries. It wasn't long before these companies realised that the lack of regulation, supervision and enforcement in developing nations, as well as their low wages and lack of workforce health and safety provisions, made them highly attractive as places to manufacture toxic chemicals or carry out polluting activities. Regulation is often rudimentary, enforcement lax and sometimes non-existent—and corruption of officialdom is rife. By choosing such places to relocate to, the chemical industry is placing itself beyond the effective reach of the law in well-regulated countries as well as exposing citizens and consumers in the developing world to a rapidly growing toxic load.

UNEP states:

> As developing countries and those in economic transition increase their economic production, related chemical releases have raised concerns over adverse human and environmental effects. Chemical contamination and waste associated with industrial sectors of importance in developing countries include pesticides from agricultural runoff; heavy metals associated with cement production; dioxin associated with electronics recycling; mercury and other heavy metals associated with mining and coal combustion; butyl tins, heavy metals, and asbestos released during ship breaking; heavy metals associated with tanneries; mutagenic dyes, heavy metals and other pollutants associated with textile production; and toxic metals, solvents, polymers, and flame retardants used in electronics manufacturing.[11]

Evidence that such effects are not confined to developing and newly industrialising countries but are rebounding on the developed world was revealed in a *Four Corners* television program made by the Australian Broadcasting Corporation, aired in July 2013. Investigating the poisoning of agricultural workers with dioxins in the 1970s and '80s, the ABC's reporters unearthed new evidence from both university and government scientists that dioxin, one of the 'dirty dozen' substances banned a decade or more earlier, was still on sale in Australia. As journalist Janine Cohen relates:

> *Four Corners* decided to test for dioxin contamination in one of the many imported products. We purchased 24D weedkiller from this warehouse in Sunshine just outside of Melbourne and sent it to government laboratories for testing. The 24D weedkiller Aminoz

625 was imported from China by an Australian company Sanonda, based in an office in Melbourne. We purchased the 24D weedkiller, which is often sold direct to farmers, no questions asked. It's almost three weeks on and we have just received our results from the government laboratory. Alarmingly, our sample of 24D weedkiller came back with dioxin levels almost seven times higher than those found by the Queensland scientists. And what's even more disturbing is we don't know how many contaminated 24D products are out there in the community because authorities are not routinely testing. *Four Corners* gave the results to Associate Professor Caroline Gaus [of the National Research Centre for Environmental Toxicology at the University of Queensland] to analyse.

ASSOC. PROF. CAROLINE GAUS: I was actually surprised because you only analysed one formulation and to actually return such a high result, I thought it was unlikely today—but again that is a reality check. When you think back to our previous study, when we actually didn't expect any contamination in the pesticides, it just demonstrates again that what we are seeing today is equally or even worse than 10 to 20 years ago, and that is of concern of course.

JANINE COHEN: Each year more than $100 million worth of 24D products are sold in Australia. And we don't know which ones contain imported ingredients.[12]

In a recent book *Bowing to Beijing*, American journalist Brett Decker and co-author William Triplett claimed:

The Chinese have peddled numerous toxic products to American consumers, including everything from children's toys to adult

> vitamins to pet food. The U.S. government regularly stops more poisonous or faulty products at the border that were imported from the PRC than from any other nation. In April 2011, for example, the Food and Drug Administration issued 197 import refusals for Chinese goods, compared to 107 for India and 105 for Mexico, the two next most prolific purveyors of bad merchandise. Some of the 197 goods refused for entry into America included hazardous cardiograph machines, cosmetics, pet medicine, diet drugs, orthodontic parts, surgical bandages, frozen spinach, asparagus and candy.[13]

What the authors omitted to note was that part of this was down to migration of the global (and US) chemical manufacturing industry into China—and that American citizens generated the demand for such imports.

During the 2020s, the newly industrialising world will become the main seat of global chemical power. Even if the *will* exists in every nation to curb the tsunami of new and untested substances or old and known toxics being unleashed on humanity, the *means* for doing so will not—at least not for the foreseeable future. Other ways must be sought to stem the flood.

Setting standards

Alarmed at the mounting public backlash provoked by the worldwide series of chemical incidents, many chemists and their organisations are espousing higher standards.

The American Chemical Association, for example, advocates a code which calls for the industry:

- To lead in ethical ways that increasingly benefit society, the economy and the environment.
- To design and develop products that can be manufactured, transported, used and disposed of—or recycled—safely.
- To work with customers, carriers, suppliers, distributors and contractors to foster the safe and secure use, transport and disposal of chemicals, and to provide hazard and risk information that can be accessed and applied in their operations and products.
- To design and operate facilities in a safe, secure and environmentally sound manner.[14]

Britain's Royal Society of Chemistry operates a somewhat ambivalent Code of Conduct for its members, in which it says, among other statements:

> All members have responsibilities arising from their duty to serve the public interest, and should be concerned with the progress of the chemical sciences. The RSC does not condone any attempt to coerce its members into refraining from lawful activity.
>
> The RSC expects members to use their professional skills to:
>
> - Advance the welfare of society, particularly in the fields of health, safety and the environment.
> - Advocate suitable precautions against possible harmful side-effects of science and technology.
> - Identify the risks of scientific activities, and take an active interest in safety throughout their organisations.
> - Undertake any lawful scientific activity as required even if in an area which arouses adverse publicity.

- Use their knowledge and experience for the protection and improvement of the environment.[15]

In the fourth point, the RSC clearly expects its members to shut up and do what the boss tells them, so long as it is 'lawful', even if they or the public are concerned over a possible chemical impact affecting the welfare of society—clearly a position it needs to rethink from an ethical standpoint.

Australia's Royal Australian Chemical Institute (RACI) is clearer in its code of ethics about where the first duty of a chemist lies: 'a member shall endeavour to advance the honour, integrity and dignity of the profession of chemistry. However, notwithstanding this or any other by-law, the responsibility for the welfare, health and safety of the community shall at all times take precedence . . .'[16]

Globally, the chemical industry operates a program known as 'Responsible Care', which is subscribed to by almost sixty countries, in which it pledges to:

- Continuously improve the environmental, health, safety and security knowledge and performance of technologies, processes and products over their life cycles so as to avoid harm to people and the environment.
- Use resources efficiently and minimise waste.
- Report openly on performance, achievements and shortcomings.
- Listen, engage and work with people to understand and address their concerns and expectations.
- Cooperate with governments and organisations in the development and implementation of effective regulations and standards, and to meet or go beyond them.

- Provide help and advice to foster the responsible management of chemicals by all those who manage and use them along the product chain.[17]

Such codes and guidelines, it should be noted, are applied only by legitimate, well-run chemical enterprises and ethical individuals in well-regulated countries—they do not govern the behaviour of chemical firms in the other two-thirds of the world, nor can they curtail the activities and behaviour of criminals, chemical weapons makers, nor miners and mineral processors, the electronics sector, the food industry or the energy, transport and building sectors. In short, such codes and guidelines apply only to a small fraction of the sources of man-made chemicals to which we are exposed.

Regrettably, even in this restricted set of cases, the pattern of global development revealed by the UNEP suggests that large parts of the chemical industry itself are progressively relocating away from countries where such high standards of behaviour—and high costs—apply. It appears that many companies which are willing to observe such standards in their home countries are hedging their bets by operating subsidiaries in more loosely governed places. At the same time it is evident that polluting industries maintain subtle and unrelenting political and economic pressure on governments and agencies in their homelands to relax both regulation and scrutiny and to facilitate approval of new substances or the reuse of old substances—frequently known toxins—in new applications.[18]

Indeed, a leading environmental guardian, the US EPA, candidly admits on its website that one of its aims is 'to make new chemicals safer, available faster, and at lower cost'—notwithstanding the potential for conflict with its stated mission, which is 'to protect human health and the environment'.[19] Although it offers valuable help to companies to clean up their act by providing access to sophisticated risk-screening computer programs, if an agency as dedicated and expert as the US EPA feels compelled to *assist and accelerate* the chemicalisation of the planet and all its inhabitants, then how much protection can humanity realistically hope to find from equivalent agencies in developing or newly industrialised countries whose products are already sold within our borders?

The task of current international regulation of dangerous chemicals may in a sense be compared with an attempt to regulate traffic flow in a large city one vehicle at a time, instead of requiring road rules and traffic lights to be observed by all. Many years ago, when the global number of chemicals was small and their dissemination mainly local, having regulations specific to certain substances in certain areas may have worked; in today's globalised economy and with the worldwide dispersal of tens of thousands of man-made substances throughout the Earth system, it has no chance of succeeding.

'There is increasing recognition among governments, non-governmental organizations and the public that human health and the environment are being compromised by the current arrangements for managing chemicals and hazardous wastes,' comments UNEP. 'These concerns take on a new level of urgency as the quantity and range of new and existing chemicals grow rapidly in developing countries and economies in transition.'[20]

The failure of rules

Man-made chemicals are so widespread in the world today because they, or their products, are very useful, very valuable and help to save and enhance millions of lives. They are a central element in the modern economy. They are never going to be universally banned—and nor should they be.

But the magnitude of our chemical—especially toxic chemical—exposure has crept up on the human population unawares. Even the chemicals industry itself, with its particular focus on specific substances and products, appears to have little grasp of the current planet-wide and pandemic impact of its products and their hidden interactions with substances from other industries and with all life forms; other polluting industries (such as mining, food and energy production) make virtually no attempt to acknowledge the pollution they cause.

This tendency to view Earth in terms of one's own locality and business interests may have been been appropriate in the twentieth century, but in the twenty-first, with the human population soaring to ten or eleven billion and a doubling in demand for all goods and services, such a view is now vastly deficient.

While continued regulation of chemicals is essential, it is already clear that regulation alone cannot prevent the poisoning of a planet, nor of the human species. Put simply, the reasons are:

- Regulation has so far banned only 0.01 per cent of the legal production of all intentionally made dangerous chemicals—in some countries. Regulation has not prevented illicit production and use, or their persistence in the environment (though it may have reduced it somewhat).

- Regulation has little prospect of restricting the production or use of tens of thousands of substances whose health thresholds are still either unknown or poorly defined.
- In all likelihood, regulation will struggle to deal effectively with the problem of chemical mixtures (as innocuous substances may have toxic effects in combination) especially in the diet, home, workplace and city environment. Also, making rules cannot prevent man-made chemicals themselves from forming adventitious new compounds as they interact with other substances, both man-made and natural.
- Emitting industries are rapidly transferring their base of production from well-regulated developed countries to weakly or un-regulated countries in order to avoid the associated costs and controls.
- Regulation has so far been unable to require the thorough safety and health testing of each new chemical and its by-products prior to commercial release, even in the most advanced societies.
- Regulation may limit, but cannot prevent, the epigenetic damage to subsequent human generations inflicted by events that take place in this generation.
- The restriction or banning of chemicals one or several at a time will, at current rates, take millennia, in addition to incurring great expense. Past experience suggests that many industries will fight tooth and nail to protect their profits and avoid the need to change; as a result the rate of progress will be glacial.
- Regulation has not succeeded in banning—or even limiting—the growth of illegal chemical production and contamination

of the environment, water and food chain by organised crime worldwide. Indeed, production of recreational drugs and 'performance enhancers' appears to be increasing. Furthermore, regulation has so far not even prevented the contamination of the environment by legal medical drugs.

In 2002, the Earth Summit adopted the 'Johannesburg Plan', which had at its heart a vision that, by 2020, chemicals will be 'used and produced in ways that lead to the minimization of significant adverse effects on human health and the environment'.[21] The prospects for meeting this goal appear bleak. It is clear that national and international regulation alone cannot restrain man-made chemical output or the creation of new substances with unknown impacts. It is also evident that it cannot even 'minimize' its effects, whatever that may mean. At best, such plans will serve to keep the issue in the public mind, encourage stricter controls in the best-run countries and set standards which others may eventually follow.

Who is responsible?

The message of this chapter is to underline that while regulation and industry codes of conduct are essential, laudable and should go much further, they *alone* cannot restrain the chemical behemoth that has been unleashed by a global economy, nor can regulation prevent the beast from likely tripling in size by mid-century. And no matter the legal outcome, nor the publicity generated by case-by-case conflicts between corporations and their victims, the outcomes of these stoushes cannot prevent chemical catastrophe—nor can the new rules they give rise to locally.

It is also time to recognise that blaming industry over individual events and demanding tougher laws will not solve the big problem either. Indeed, blaming industry is likely only to harden its resolve to become less transparent and fight fiercer rear-guard actions—and will simply encourage manufacturers to move to less strictly regulated places where the government is more pliable or sympathetic. It is far, far better that a contaminating industry be based in countries where its own standards—and government standards—are high, rather than move to a place where it will escape scrutiny yet continue to compound global contamination.

It is time for everyone to realise that we ourselves, by virtue of the demands we place on industry for ever higher living standards and cheaper products, are complicit in the poisoning of the Earth system by human chemical activity. Such poisoning is an outgrowth and inevitable consequence of our own wishes, wants, needs, whims and fashions expressed in the consumer society, which the market is trying to satisfy. It is *our* desire for safety, comfort, convenience—and, above all, for cheaper products—that drives the market and creates the monetary incentives for chemicals to be made. As Mahatma Gandhi once said 'The future depends on what you do today': we need to appreciate that our demands come at a high price, and that price is increasingly personal. If we use toxins against insects or weeds, or contaminate our homes, groundwater and soils, consume food produced with chemicals, cosmetics, drugs, electronics and manufactured goods, if we use fossil fuels, minerals and electricity, then we unavoidably end up contaminating ourselves, our children and everyone else into the future.

On a crowded planet, every act of consumption has chemical consequences. Some of these are lethal. It is time we understand this, at the level of the species as well as the individual. It is a matter of collective, as well as personal, responsibility.

Man-made chemicals are directly responsible for killing around five million people per year, and injuring eighty-six million. Indirectly they kill and afflict tens of millions more with a burden of disabling disease that grows with each passing year. There is mounting scientific evidence that they affect everyone, everywhere. We are all actors and agents in this process, even though we may not care to admit it or are yet able to face up to the responsibility it carries. Yet, in our hearts, every one of us who reflects on this will know it to be true.

In a sense, we are *all* getting away with murder.

CHAPTER 8
CLEAN UP SOCIETY

The pandemic . . . caused by industrial chemicals is,
in theory, preventable.

Professor Philippe Grandjean, Harvard School of Public Health

'And the winner is . . .'

On 23 September 1993, thousands of Australians jumped for joy as Olympic Games boss Juan Antonio Samaranch proclaimed, 'and the winner is . . . Sydney'.[1] Unbeknownst to the hundreds of thousands of athletes and visitors who flocked to the Sydney Olympics in 2000, the words also triggered one of the largest and most successful environmental clean-up efforts on record.

Beneath what became the Olympic site and village, quietly festering away, there lay an estimated nine million cubic metres of waste and contaminated soils spread over 400 hectares of the 760 hectare site. A toxic outgrowth of the 'consumer society', this garbage had accumulated rapidly since the 1950s and included petroleum waste, unexploded ordnance from an old military store, acid sulphate soils, illegally dumped wastes along the waterways (including persistent organic pollutants, polycyclic aromatic hydrocarbons etc.), dredged sediments, municipal waste in managed tips, industrial waste (such as rubble, power station fly ash, gasworks waste and asbestos) and contamination from burning pits, chemical leaks and chemical use. In terms of what it was chucking away and jettisoning from memory, Sydney was no different to any other big industrial city around the planet. All metropolises have hideously toxic things underfoot, about which we prefer to be ignorant—but which long remain a sleeping, seeping menace.

The clean-up of this industrial filth became the largest project of its kind in Australia and remains one of the most significant positive environmental legacies of the Sydney 2000 Olympic and Paralympic Games. The site was extensively investigated using exploratory boreholes and a remediation plan adopted which involved safely containing and, where possible, treating waste on site, rather than relocating it to other places. Nine million cubic metres of waste were recovered, consolidated and relocated to designated waste containment mounds. These were capped, sealed, landscaped and turned into parkland. Collection and transfer systems were built to prevent chemicals leaching into the environment. Four hundred tonnes of soil contaminated with hydrocarbons and hazardous chemical waste were treated in a two-stage thermal desorption and destruction process. Dust, vapours, noise and water were all continually monitored.[2] Nearby Homebush Bay, previously regarded as 'one of the most contaminated waterways in the world, due to the cocktail of organic contaminants', including DDT, dioxins and furans escaping from a former Union Carbide plant,[3] was so heavily polluted that authorities banned residents from eating the local fish became, pre-Olympics, the target of a major clean-up effort.

When it was completed, the Sydney Olympic Park was about as clean, safe and healthy as the technology of the day could render it. Besides proving it possible to make safe some of the world's most horribly polluted land, a major benefit was the dramatic gain in local land values—from being next to worthless, the remediated land rapidly acquired inner city property values and attracted contracts for major housing, recreational and retail developments. The clean-up operation cost $137 million, but the renovated land ultimately yielded a return of around $600 million.[4] Not many investments generate such an attractive rate of return so quickly. This illustrates the economic gains to be had from cleaning up yesterday's (and today's) urban messes, not to mention the long-term costs of healthcare and premature death that the clean-up may have averted. The lesson from the Sydney Olympics is quite simply that a cleaner world is a more prosperous and profitable world—as well as a safer one.

Going global

Compared with the Olympic Park clean-up operation, the challenge of reducing the individual and planet-wide burden of toxic man-made chemicals is very large—it may at first sight appear to be an intractable and insurmountable problem—but with global awareness, good technology, partnership and a clear conviction of the need to care for ourselves, our children and our Earth, it is possible to overcome it. Together.

There is little that humans have done that cannot be improved, rectified, replaced or undone. The challenge is to build a worldwide consensus for rapid and concerted action that makes these things happen.

The first point for all citizens to understand, however, is that Earth-system chemicalisation will not be solved if left to government regulation and industry compliance alone—even when they act from the best of motives—so long as billions of consumers all around the world persist in sending out the economic commands that stimulate the mass release of toxins. And while it is a global problem, it cannot be solved by enforcement at global level, because international agencies have little or no power to act at national, local and industry levels—however, global agencies do have the ability to influence, guide, inform and provide templates for region-, continent- or planet-wide action. Because, like it or not, we citizens of the world are all in this together: even if individual countries are willing to reduce their own toxic output and use, they and their citizens cannot escape the worldwide influx of contaminants in air, water, food, wildlife, imported products and damaged human genes.

Eugene Smith took his searing images of Minamata and published his book because he hoped that, through awareness, such evidence might provide an indelible lesson, to forever guard against any recurrence. Yet, for all the rising tide of public awareness, cases of chemical poisoning have happened time and again and are still happening, on an ever-increasing and now virtually universal scale. By all indications the poisoning will continue to escalate for the remainder of history until our health and that of our offspring are damaged beyond repair.

Unless something profound changes. Before we address what that change entails, let us review briefly the many very encouraging methods currently available to clean up our communities, our industries, our countries and the planet, and the main instruments which make it possible.

Legal clean-up

Many countries around the world now have well-established regulatory structures and expert environmental agencies for keeping track of chemicals in their various forms—as commodities, constituents of products, environmental pollutants, occupational and public health hazards and wastes—and regulating their use. These agencies share chemical knowledge among themselves, making the task of the identification of toxins less onerous and costly than it would be for each individual body. They also set standards of excellence which others seek to follow. They perform an invaluable educative role, but one that tends to be restricted to an audience of informed public; their publications are seldom easy for people without science training to understand. However, these agencies are also frequently nervous about antagonising industry or the politicians it controls. Environmental protection agencies are becoming stronger and more expert, but have so far been unable to keep pace with the tide of new chemical releases, safety-testing of these new chemicals, and the issue of how to deal with chronic low-level poisoning of their populations from a multitude of different chemical mixtures originating from various sources. Often, too, such agencies are depicted by their industry critics as creating 'needless' barriers to business or employment growth, and they are thus understandably cautious about any action they may pursue. This makes them sometimes reluctant to act on a purely precautionary basis, as citizens might wish.

In some countries, the chemical, food and other manufacturing industries have self-imposed safety and health standards and ethical guidelines. These tend to apply to particular products and processes, and not to the combined chemical burden on the human population or the food chain or the mixtures our food contains. Also, the policing of these standards is seldom transparent to the public. Furthermore such codes of practice are far less common in industries such as mining, energy production, transport, construction, agriculture and so on, all of which are major contributors to the release of man-made chemicals into the global environment.

The United Nations' Strategic Approach to International Chemicals Management (SAICM)[5] and the various chemical-related multilateral agreements (see Table 3) provide a range of voluntary and legally binding frameworks designed to encourage better management of chemicals. SAICM was formed in 2006 to give impetus to the Johannesburg Declaration on Sustainable Development of 2002[6] and has as its declared aim that, by 2020, chemicals should be 'produced and used in ways that minimize significant adverse impacts on human health and the environment'.

'Acknowledgement of the essential economic role of chemicals and their contribution to improved living standards needs to be balanced with recognition of potential costs. These include the chemical industry's heavy use of water and energy and the potential adverse impacts of chemicals on the environment and human health. The diversity and potential severity of such impacts

TABLE 3 **Key international chemical treaties and agreements**

- Aarthus Convention on Access to Information, Public Participation in Decision-Making and Access to Justice in Environmental Matters 1998
- Basel Convention on the Control of Transboundary Movements of Hazardous Wastes and their Disposal 1992
- Convention for the Protection of the Marine Environment of the North-East Atlantic (OSPAR Convention) 1992
- Convention on the Ban on the Import into Africa and the Control of Transboundary Movement and Management of Hazardous Wastes within Africa (Bamako Convention) 1991
- Convention on Long-Range Transboundary Air Pollution (LRTAP Convention) 1983
- Convention on the Prevention of Dumping of Wastes and Other Matter (London Convention) 1972
- Convention to Ban the Importation into Forum Countries of Hazardous Waste and to Control the Transboundary Movement and Management of Hazardous Wastes within the Pacific Region (Waigani Convention) 1995
- International Convention on Oil Pollution Preparedness, Response and Cooperation (OPRC Convention) 1995
- Montreal Protocol on Substances that Deplete the Ozone Layer 1989
- Rotterdam Convention on the Prior Informed Consent Procedure for Certain Hazardous Chemicals and Pesticides in International Trade 2004
- Stockholm Convention on Persistent Organic Pollutants 2004

Source: US EPA, <www.epa.gov/oswer/international/factsheets/200610-international-chemical-hazards.htm>.

makes sound chemicals management a key cross-cutting issue for sustainable development,' says SAICM.

However, all these international instruments operate by suasion and have little actual power to ban chemicals or impose penalties for their use—that is left to government agencies in individual countries, which are often under industry pressure to do as little as possible. Also these international agreements tend mainly to focus on individual substances and pathways which have been shown to cause harm, rather than on the central issue of the rising chronic exposure and combined chemical burden of all humanity.

Thus, overall progress has been slow. So far only twenty-one out of an estimated 40,000 suspected toxic substances have actually been recommended to be banned or severely restricted internationally over a decade, and this means that new chemicals are being introduced into the Earth system at a rate around five hundred times faster than old ones are being withdrawn.

Says UNEP:

> The limited attention to chemical safety in national planning derives in part from the lack of coherent risk reduction strategies among the various government authorities responsible for chemicals and wastes. To develop more coherent risk management strategies, cross sector coordination is needed on chemical management among these agencies. In addition, it is important to ensure clear roles for both government and the private sector. In many countries corporations have good information on chemicals and wastes management and the technical capacity to launch effective strategies. Many global corporations actively propagate effective chemical strategies and techniques along their value chain and within related industries.

> By making chemicals and product manufacturers and importers the first line of sound chemicals management, the responsibility and costs for social and economic development are more effectively shared between private and public sectors.[7]

UNEP and others advocate a 'multi-stakeholder' approach to dealing with chemicals which brings together government, industry and civil society in an effort to reduce the human chemical burden. However, while industry and government are sometimes in alignment, the voices of citizens and consumers are far more muted and uncoordinated; in many countries, they remain largely unheard. Building even a national consensus for action is painfully slow—and in the meantime there is no escape from the rising tide of contamination reaching us from places where such a consensus may never be attained.

Clean-up approaches

Under the leadership of chemists with a conscience, enlightened industry in many countries is working on approaches which show great promise in assisting to reduce the overall chemical burden on humanity, especially if supported by good regulation and market support, as well as approval from consumers and citizens. These approaches include:

Life cycle assessment (LCA): examines all stages of a product's life from cradle to grave (or cradle-to-cradle as it is sometimes known)—from the original extraction and processing of raw materials through manufacture, distribution, use, repair and maintenance, and disposal or recycling—with a view to minimising pollution and waste at every stage. Besides the chemicals which comprise the object, this assessment includes the energy, carbon and water used in making or distributing it. The advantage to industry is that LCA can also save money; however, it is expensive to set up.[8]

Material flow analysis (MFA): is used to study the route of material flowing into recycling or disposal sites, and stocks of materials, in space and time. It links the sources, pathways, intermediate and final destinations of the material. It is used to reduce waste, energy use, unwanted by-products, costs and contamination all along the chain.[9]

Multi-criteria analysis (MCA): a computational approach that helps to solve complex problems in areas such as production and waste disposal, and to minimise volumes of waste and contamination.[10]

Extended producer responsibility (EPR) or 'product stewardship': an approach that requires the producer to take back the product after its useful life is over and to then recycle it safely. EPR is already commonly used for products such as glass bottles, aluminium cans and printer cartridges; however, there is potential application to a great many products including computers and mobile phones. The virtues of EPR are that it extends the life of scarce resources including metals, minerals and energy and also reduces the amount of pollution generated in subsequent generations of the product.[11]

Green chemistry: a philosophy of chemical production that encourages the design of products and processes that eliminate the use and generation of substances harmful to humans, nature and the environment. For example, it involves creating 'soft chemicals' such as plastics, fuels, pesticides, oils and solvents by biological methods from farm crops or algae instead of from petroleum or coal. Driven by the quest to cut their own waste and reduce use of energy and water, many of the world's big chemical firms are showing interest in this approach. The global 'green chemical' market is projected to grow to $100 billion by 2020 (UNEP). Despite such promise, 'green chemical' patents still make up less than 0.01 per cent of all new chemical patents.[12]

Green manufacturing: means designing and producing products—usually by lifecycle analysis or a similar method—that generate far less waste at every stage of their life and which then can be successfully recycled without leaving a toxic footprint. It involves the idea of 'resource productivity': making more with less.[13]

Green building: means creating and operating buildings and structures that use far less raw materials and energy, via the use of lifecycle analysis and planned recycling of used or waste materials. This involves reducing or eliminating the chemicals released in raw material extraction, transport, processing and building operation, demolition and re-use of materials.[14]

Integrated pest management: is a popular concept being widely adopted in agriculture to reduce the use of pesticides on the farm and in the food chain by managing crops and food commodities in ways that make them less prone to diseases, weeds, fungal contamination or pest attack. It includes the use of enhanced crop varieties and mixtures, rotations, and approaches to pest control such as the introduction of beneficial insects and companion planting or trap crops to divert them.[15]

Organic farming: uses traditional farming methods such as crop rotation, composting and natural forms of pest control to eliminate or reduce use of synthetic pesticides, fertilisers and livestock medications and produce healthy soils. However, in developed countries it still involves considerable use of fossil fuel, rock phosphate and other materials.

Chemical leasing: a business model developed in the paint industry where a company does not sell, but rather leases the chemical to a subcontractor to carry out a specific service. This tends to promote a culture of reduced chemical use to get the job done, rather than industry being driven by attempts to increase the volume of chemicals sold and used.

Industrial ecology: put simply, the idea is to locate various waste-producing industries close together so that what one throws away may be used by another industry as a feedstock. The aim is to reduce all forms of waste and find productive uses for it. Again, it involves the study of material and energy flows, and also seeking the best ways to make use of these flows by co-locating industries or productive activities that can benefit from one another.[16]

Zero waste: a series of principles embracing the use and reuse of all products. Zero waste mimics cycles in nature, where nothing is wasted. It involves designing and managing products and processes to systematically avoid and eliminate the volume and toxicity of waste and materials, to conserve and recover all resources, rather than burning or burying them. Implementing zero waste seeks to eliminate all discharges to land, water or air that are a threat to planetary, human, animal or plant health.[17]

Risk assessment and remediation: as shown in the Sydney Olympics example, current and historically contaminated sites can be quite effectively cleaned up using a range of advanced techniques. The main obstacle to their widespread use is cost—and the reluctance or inability of some industries and governments to pay for clean-up—especially in newly industrialising countries such as India and China. Technical difficulties arise in cases where there is a complex mixture of many toxic compounds. Nevertheless, it is in theory feasible to assess the risks posed to human health by each of the world's estimated five million contaminated sites one by one—and to clean them all up where the risk is judged too great. Risk assessment thus puts the focus of the clean-up effort on the most hazardous sites and substances and reduces costs by avoiding remedial work in cases where contamination is not likely to lead to harm. Typical methods for cleaning up toxic sites include:

- 'dig and dump': the dredging or excavation of toxic material and its removal to a safe landfill
- 'pump and treat': the process of pumping contaminated ground water and its cleansing by various methods
- stabilisation: adding special chemicals to contaminated material to stabilise or break it down
- solidification: use of special chemicals to solidify contaminated material to prevent it from dissolving or moving
- in situ oxidation: injection of oxidants to break down susceptible contaminants in groundwater and soil
- bioremediation: use of special microbes to break down or take up contaminants
- phytoremediation: use of special plants to absorb toxic metals and other substances from soil or water for removal and safe disposal or re-use

- electrokinetic remediation: involves the removal of heavy metals from contaminated soils by attracting them to the poles of an electrical current passed through the soil
- heat treatment to break down toxics
- ultraviolet light treatment
- permeable reactive barriers: an underground barrier built across a groundwater flow that absorbs or breaks down contaminants in the water—chemically, biologically or both.

As illustrated by the case of the Sydney Olympics clean-up, it is quite feasible to render a badly contaminated site reasonably clean and safe using today's advanced remediation methods. Examples of other celebrated clean-ups are described below.

Safer Singapore

The island state of Singapore has a commitment to cleaning up its environment dating back to a tree-planting campaign in the 1960s and the cleansing of the once famously foul Singapore River in the 1980s.[18] Today, one of its major challenges is waste disposal, and with just one landfill site to serve the state's five million inhabitants—who generate around seven million tonnes of garbage a year—recycling is high on the public policy agenda. Singapore manages to re-use a remarkable four million tonnes of the discarded waste, especially e-waste. The cornerstone of Singapore's waste management is a campaign to engage the community in recycling, with success rates varying from as low as 10 per cent for plastics and 12 per cent for organic waste to as high as 96 per cent for metals and 99 per cent for building materials. Non-recyclable waste is mainly dealt with by incineration (which supplements the city's electricity) and the remaining waste is disposed of in a 'landfill' on the island of Semakau which has been reclaimed from the sea by a large enclosing dyke. With over 2000 companies using toxic chemicals in Singapore, their use and disposal is tightly monitored with, again, a remarkable quantity being recycled and the rest treated in a special facility and their residue sent to landfill or used as road base.

Superfund success

Over the past three decades, says the US Environmental Protection Agency, its Superfund scheme has located and analysed more than 91,000 hazardous waste sites, protected people and the environment from contamination at the worst sites, and partnered with states, local communities, and others in clean-up operations. It maintains a list of national priority sites which it ranks for clean-up priority according to:

- the likelihood that a site has released or has the potential to release hazardous substances into the environment
- the toxicity and volume of the waste
- whether people or sensitive environments may be affected by the release.

A pleasing feature of Superfund clean-ups is close consultation with affected communities, which involves listening to their concerns and what they want done about the issue. As of mid-2013, more than 1300 contaminated sites had been fully cleaned up in the US under the Superfund scheme, an average of around forty sites per year. The clean-ups were funded by the taxpayer, but with substantial private contributions under the 'polluter pays' principle, in cases where a polluter was still in business.[19] The US EPA publishes an extensive list of treated sites.[20]

All these examples provide a brief overview of some of the practical methods which have been developed to reduce the burden of man-made chemical contamination on the planet and all its inhabitants. Many industries, companies, scientists and government agencies support such activities and if these actions were to be universally adopted, there is no doubt they can make a major difference to the global chemical burden. It must be said, however, that the people and companies who support such concepts are in the minority in most societies, in countries both developed and developing. Furthermore, while there has been noteworthy success in cleaning up particular sites, those cleaned up remain a relatively small proportion of the estimated five million contaminated sites thought to exist worldwide, and increasing by the year. Also, due largely to the migration of the less scrupulous industries to the poorly regulated regions of the world—and due to rising consumer demand driven by world economic growth and globalisation—the flood of contaminants continues to grow by leaps and bounds.

In other words there is insufficient reward for companies, industries and countries, for doing the right thing by humanity, posterity and the Earth. For this and many other reasons, there is no doubt that it is far better (as well as less costly) to prevent a toxic mess from occurring than trying to clean up its aftermath.

Professional shortcomings

An issue which cannot pass unremarked is the fact that public awareness of ill-health resulting from chemical pollution is low, in part because medical awareness is also low. Indeed an irony of modern medicine is that it is now largely dependent on petrochemicals (euphemistically known as medications) to provide 'cures' for conditions which, in many cases, may originate with or be mediated by the global human chemical burden. Illustrating this is the fact that a number of common anti-cancer drugs are themselves known to cause cancer.[21] An important issue raised by Dr James Siow of Australia's Institute of Integrative Medicine is the reduced exposure of today's medical students to toxicology training in their medical courses which, he says, makes them less equipped than their predecessors to recognise and correctly diagnose conditions attributable to chronic chemical exposure when they see such cases in their consulting room. Instead, there is a general trend for doctors to prefer explanations in terms of the patient's own genetics or mental state, which then leads to further chemical prescription.

Behind this issue is the well-documented concern that many Western doctors are being actively trained, encouraged or induced by pharmaceutical companies to prescribe more chemicals for an already over-chemicalised race of humans. As Ray Moynihan and David Henry described the issue at the world's first conference on 'disease mongering', this is: 'the selling of sickness that widens the boundaries of illness and grows the markets for those who sell and deliver treatments. It is exemplified most explicitly by many pharmaceutical industry–funded disease-awareness campaigns—more often designed to sell drugs than to illuminate or to inform or educate about the prevention of illness or the maintenance of health.'[22] An aspect of this, referred to in Chapter 3, is the rapid growth in the number of deaths worldwide caused by antibiotic-resistant bacteria which, in turn, is partly due to over-reliance by modern medicine on chemical solutions and the overprescribing of antibiotics: 'up to 50% of all the antibiotics prescribed for people are not needed or are not optimally effective as prescribed,' comments America's CDC.[23]

Large corporations also sometimes distort the science around the causes of disease. This is graphically illustrated by the tobacco industry—which is claimed to have spent $370 million over a decade on genetic research intended to persuade the public that lung cancer was due more to the victim's own genes, than to smoking.[24]

And finally, no doctor or pharmaceutical company can truthfully answer the question: how will this prescription drug interact with all the thousands of other toxic chemicals which my patients already carry in their bodies, and are being exposed to daily? This question has not been scientifically fully explored. While we know a fair bit about specific side-effects of particular medications and their interactions with some other drugs, foods or drinks (e.g. alcohol), we know little about medications' wider interactions with the complex chemical mixtures found in the human environment and food chain. All we can say at present is that the ingestion of chemical medicine adds to, and may increase, the overall toxicity of the human body burden, as is often the case in chemical mixtures. But we do not know the risks of such chemical combinations.

All these medical shortcomings—the lack of toxicological knowledge, the lack of adequate research, the tendency to bow to pharmaceutical companies and their occasional attempts to bias the science in order to sell more product—render the task of preventing disease in the human population far more difficult, especially in cases of diseases related to chemical exposure. If doctors the world over still take seriously their Hippocratic injunction, '*Primum non nocere*' ('First, do no harm'), then it is time for major reform of the ways Western medicine is both taught and practised around the world, aiming towards the clear goals of firstly reducing the human body burden of chemicals—regardless of whether the exposure was accidental or prescribed—and secondly, preventing rather than merely treating disease. For example, since medications are implicated in about half of the medical mishaps estimated to kill 200,000 Americans, 18,000 Australians, 30,000 Britons and 35,000 Germans each year, a good start might be to reduce the death toll caused by chemical medicine itself.

A second profession in need of reform is chemistry. While most chemists are ethical, well-intentioned people, it is hard to deny that many thousands of chemists have also laboured over the past century to create more deadly weapons, more toxic poisons and more harmful substances, without much regard to their unintended impacts on the lives of millions of ordinary people—even their own children. Also, many chemists appear breezily inclined to under- rather than over-estimate the hazards of chemistry, and are occasionally heard to express contempt for the concerns of citizens and families who do not want to be poisoned. The Royal Chemical Society's ambivalent code of ethics makes it obvious that many chemists are conflicted over whether they owe their primary loyalty to their employer, be it corporation or government, or to humanity at large. It is essential that aspiring chemists both study and clearly comprehend their ethical responsibilities at the same time as they imbibe their professional knowledge; such tenets are already part of the training for medical professionals. Added to chemists' training must be a stronger dose of toxicology—and that means reforming how chemistry is taught in universities worldwide. Perhaps it is time that chemists, too, took an oath 'primum non nocere'. It is crucial that the chemistry profession adopt a unified stance that is committed not to the augmentation, but rather to the reduction of the chemical burden in all people and in the Earth system, as quickly as possible. Collectively, the profession should deploy all its intellect, effort and integrity in the tasks of developing low-cost, workable, green chemistry alternatives, and in tackling problems such as chemical mixtures and contamination of the Earth system as a whole, and identifying practical ways to mitigate them.

A gender issue?

Readers may have remarked, perhaps with irritation, the use of the expression 'man-made chemicals' throughout this book. The words are advisedly chosen. The vast bulk of the world's toxic emissions are created and produced by men, not by women. A study of the gender balance in the chemistry profession globally reveals it to be heavily slanted to males. For example, the US Chemical Heritage Foundation stated 'Women constitute only 16% of the tenured faculty in chemistry departments at U.S. colleges and universities.'[25] Only four out of 166 Nobel Laureates for

Chemistry created since the award was inaugurated were female.[26] A survey of women in scientific careers in the US found that in 2007 only 9 per cent of the executive officers of forty leading chemical companies were women and only 12 per cent of their board members were female.[27] While women make up more of the junior ranks of chemists, they have not been responsible for taking the big decisions about what is produced and what is not. The British film *A Chemical Imbalance* explores the gender bias which appears to exist universally in this vital profession.[28]

Women, on the other hand, are very prominent as leaders among the various organisations, consumer bodies and parents' groups calling for a reduction in the toxic burden of society and, from this, it is not a long step towards hypothesising that the chemical poisoning of humanity is as much a gender issue as violence, sexual persecution, discrimination or other forms of economic and political disadvantage. Whether it is simply that men are more 'gung ho' and inclined to underestimate the risks of chemicals—a theory which finds some support in case studies of farm workers—whereas women are more foresighted in wishing to assure a healthy environment for children to grow up in is not yet subject to proof, but seems like a good doctoral topic for someone.

Be that as it may, it appears possible that a world in which women share equal power with men would also be a much less toxic world. And a chemistry profession in which women enjoy equal status would probably be more precautionary in its approach to the issue of global chemicalisation—as well as quicker to reject harmful substances, take responsibility for its mistakes, and to both consider and adopt safe alternatives.

Can we adapt?

Humans are nothing if not adaptable, and in recent years scientists have clearly shown that we are continually evolving as our genes respond to new signals in the environment around us. A classic example of this is the physiological ability of people in cattle-rearing societies to digest lactose, while non-cattle-societies are often lactose intolerant: the former groups of people have evolved lactose tolerance to suit their diet and this has happened in the span of just the last 6000 or 7000 years when humans have herded cattle. A second example is high-altitude-dwelling people such as the Nepalese who have evolved blood chemistry which enables them to cope with low oxygen conditions much better than lowlanders. Could the same apply to the chemical shower to which we are all now subject? Could we *evolve* a greater tolerance to toxins in our food, air, water, homes and workplaces? Some scientists have wondered.

The short answer, most probably, is no. Firstly, the chemical mixtures that come at us from all sides are far too diverse and too uneven in their dosages for us to make sufficient adaptive genetic or epigenetic changes within a reasonable time frame. We might perhaps adapt to a few substances to which we are continually exposed—but not to thousands of individual substances, or to billions of mixtures. Secondly, in order to preserve any beneficial changes in our gene pool, we would have to stand by and watch the deaths of people who do not adapt (preferably before they procreate and pass on genes that are less chemically resilient) and we certainly would not do that. In this respect modern healthcare has effectively derailed evolution; in other words, we now preserve not only advantageous genetic changes, but also many disadvantageous ones in our population, and these include chemical sensitivities and susceptibilities.

In short, hoping for timely genetic adaptation of the human species to the man-made chemical overdose is highly unlikely to prove a viable strategy for our ongoing survival on a poisoned planet.

Removing the burden

Reducing both the volume and the toxicity of the chemical assault on humanity is the only answer—and achieving these reductions may not be as difficult as it might appear.

At some point in the coming decades, humanity will face the necessity to entirely cease burning all fossil fuels—coal, oil, natural gas, shale oil, tar sands and so on. This will happen when the impacts of climate change become so horrendous and inflict such constant damage on people, as well as on food supplies, homes, industries and the natural environment—thus driving mass migrations, conflict and pandemics—that an angry and terrified society is moved to demand their urgent substitution with clean energies. Owing to the entrenched stance of the fossil energy lobby and its supporters, as well as society's natural inertia and reluctance to take action for as long as any problem appears 'off in the future', such action may take a couple of decades to eventuate—but happen it will.

Besides being responsible for causing climate change, oil and coal are also the primary sources of most of the toxic chemicals of concern to humanity.[29] They provide the feedstock for the manufacture of industrial chemicals, pesticides, legal drugs, illegal drugs, plastics, solvents, chemical weapons and other pollutants, including most of the endocrine disruptors. Their elimination will go a long way towards cleansing our poisoned planet.

That human health is being undermined globally, and children and our genome damaged at this very moment by oil and coal, constitutes an argument for *immediate* action.

This means that there are now two utterly compelling reasons—our own health and that of the Earth's natural systems—to replace all forms of fossil fuel and product use with cleaner, safer, healthier, renewable solutions as quickly as possible. How we respond to these challenges will have a powerful influence on the human destiny and the fate of all life on Earth.

Furthermore, there is growing evidence that fossil fuels can be substituted with other, less harmful, forms of energy. For static energy, alternative forms include solar thermal, photovoltaics, wind, tide, wave, hydro, nuclear, 'clean' nuclear (thorium cycle), hot rocks and geothermal. These and other energy sources have long been proposed and many are catching on quickly, while for transport fuels the use of oil extracted from algae grown in salt water is emerging as a highly promising alternative and is, indeed, attracting investment from the world's leading airlines and aviation firms, the US military and more than thirty countries worldwide. Algae consist typically of about 50 per cent oil and require only sunshine, salt water, warmth and nutrients (which can be supplied by urban waste) to grow; as long as these basics are available, fresh oil can be produced continually and at prices that are increasingly competitive with fossil petroleum as the technology advances. It has been calculated that the world's entire diesel fuel requirements could be met from an area of 57.3 million hectares of saltwater ponds, or pontoons in the sea.[30] Algae can also be used to produce goods such as human health foods, livestock feed, biodegradable plastics, synthetic textiles, fine chemicals, 'soft' pesticides and many other useful things—all while helping to lower humanity's greenhouse emissions and overall chemical exposure. The pros and cons of this important new industry are discussed in the US Department of Energy's algal biofuels roadmap.[31]

In addition, there appears considerable scope to produce less toxic chemicals from plants and other biological sources by novel processes. Dr Allan Green, an eminent plant scientist with Australia's CSIRO says: 'Most work to date is on bio-based sources of the exact same molecules as petrochemicals, in which case toxicity would also be the same. However, [there is] significant opportunity to produce functional equivalents that might have different chemical structures ('bio-similars'), and potentially less harmful effects.'[32] Dr Green believes that today's main oil crops (such as oil palm, canola and soybean) will struggle to keep pace with human food demand, let alone be able to supply the chemical industry—but breakthroughs that include breeding crops that produce oil in their leaves could triple oil output.[33]

Besides carbon emissions from the burning of fossil fuels, a second very large source of chemical pollution by volume is the mining and minerals processing industry. Here the substances released are, for the most part, less toxic and more easily controlled through good technology, mine design and plant management; the key will be finding the right mix of regulation and market incentives to encourage the global adoption of best practices. The best strategy of all is to accelerate the recycling of all metals, which will dramatically reduce by-production of contaminants during mining and mineral processing. Our existing waste streams will become the mines of the future: promising high technologies are already under development around the world to separate and extract valuable metals and minerals from the materials which most of the population currently considers as 'waste'. It is very likely that at some point this century, the human population will peak; thereafter the demand for raw minerals—and for mines—will begin to decline permanently. Mineral companies that wish to remain profitable will switch their attention to winning metals from substances already mined, in preference to digging further new holes which wastes energy and creates pollution. Approaches such as phytomining and biomining—mineral extraction using plants and microbes—will aid the evolution of the resources industry, along with sophisticated high-temperature processing of metallic wastes using solar furnaces. So too will the design of manufactured products that are purpose-built for easy recycling; indeed, like the bottle recyclers of old, smart companies may even pay to get their discarded products back in order to re-extract their materials and components.

A third massive source of contamination in our lives is hazardous waste. Much of this derives in the first instance from fossil fuels, so the elimination of petroleum and coal use will automatically cleanse our hazardous waste stream substantially. Recycling, biological and advanced thermal treatments, combined with sound regulation and consumer pressure for clean products, can probably replace the remainder of hazardous waste creation—along the lines being pursued in Singapore.

A fourth huge pollution source is nutrient release, chiefly from agriculture and poorly planned development. The alternative to this is the establishment of large-scale urban agriculture and the recycling of nutrients. Cities already concentrate a large part of the world's nutrients and fresh water, and mostly throw it away as waste. Using this waste, urban farms and novel food systems (such as synthetic meat, algae farms and biocultures) are capable of supplying at least half of the world's food, without the use of chemicals or even synthetic fertilisers. These systems also have the virtue of being climate-proof, unlike traditional farming.

Once such industries become established, the remaining crop and livestock industries will be under far less economic pressure, not having to compete at the cheap end of the market. They will no longer be forced to use high-intensity chemical production systems and can turn to regenerative farming and grazing methods that heal landscapes, end soil erosion and lock up carbon. This will in turn help revive the dead spots in our lakes and oceans and bring back lost fishing industries, restore and revegetate the world's rangelands with all their wildlife, and develop more biologically complex and sustainable arable farming systems. Smart farmers everywhere are already embarking on this quest: what they need most at this point is support from smart consumers, smart supermarkets, smart food firms and smart governments.

As the Europeans have already demonstrated, three-quarters of the pesticides used in agriculture and the food chain can be eliminated without making any difference at all to food supplies or prices—other than improving their health and safety. This amounts to evidence that much of the chemical use in world agriculture is currently unnecessary and can be significantly reduced using sustainable systems for managing pest and weed damage—without risking the health of farmers, consumers or even bees and without significant yield loss in the world's crops and pastures. However, as the global farmers' market movement is demonstrating, if consumers are willing to pay farmers more to produce clean food, then chemical use in farming can in fact be dramatically reduced or even eliminated, as can the associated degradation of soil, water and rural communities.

For all this to happen, however, it is first necessary for citizens to gain an appreciation of the scale of their present chemical burden and the urgency of curbing it. Only then will we generate the economic and market signals that industry, including agriculture, needs to move to new, cleaner and safer methods of production and products. And only then will governments heed the wishes of the new majority, who will no longer stand by while they are poisoned. So, how is this to happen?

Taking responsibility

An essential first step in overcoming the contamination of the Earth system is for us to understand and accept our own personal responsibility in it. In one way or another—but most significantly by the demand signals we send when we purchase food, goods and services—each of us encourages the production of toxic substances throughout the world, whether we know it or not. We want things to be cheap, convenient, safe and useful; while chemicals help to achieve that, we have been blind to the risks their use entails. Governments and industry, too, prefer 'ignorance', which is why most new chemicals are rushed into global release without proper safety testing. As a society we have naively swallowed the narrow view that there is a chemical solution to most of our problems. Such a view tends to promote the upsides and to ignore possible downsides. We have gratefully accepted the blessings of chemistry without inquiring too deeply into its hazards—and without even wanting to know. So, to start with, greater knowledge will help us to change the nature of production.

Building on this first step—the acknowledgement of our shared responsibility for the chemical pollution of ourselves and the planet—moves us beyond the sterile and unproductive argument over who is to blame, and whether or not industry is at fault for producing what we, the market, order it to produce by our spending patterns. We have to understand that if we demand cheap food, clothes or furniture, then chemicals are an almost inevitable accompaniment, because industry will be driven to produce by the cheapest possible methods, and they are usually chemical.

Thus, what we save at the supermarket or department store, we spend at the pharmacy, hospital and hospice—or on caring for our chemically damaged children. Considered from such a perspective, might it not be wiser to spend just a little more on our food and other consumer goods, thus giving industry the incentive to produce them by less toxic methods?

The task before us is to provide enough consumers with evidence-based information at a global level to drive change in industry and government by sending the right market signals and economic rewards to industry for doing the right thing. Indeed, the market is probably the *only* way we can discipline or restrict industry in ill-regulated regions of the world—that means choosing not to buy its toxic products and preferring clean ones produced by ethical industries that are sensitive to public wishes and transparent about their production systems.

It's not up to industry or government to lead the change. A cleaner world is up to us.

CHAPTER 9
CLEAN UP THE EARTH

Nothing is more important than your children's health.
We help you protect them from the toxic exposures
they face every day.

Nancy and James Chuda

'Colette was our daughter. It is because of Colette—and the countless other children whose lives are being lost to cancer and other childhood diseases that are linked to hazards in the environment—that we are reaching out to you. We want to share with you what parents who have lost children tragically to cancer have learned, and what together we can do about this terrible threat to all children,' say Californian couple Nancy and James Chuda on their website, *Healthy Child Healthy World*.[1]

The Chudas founded their non-profit public advice website after the loss of Colette to a rare cancer, Wilms' tumour, with the aim of helping parents make better product and lifestyle choices in order to protect their children's health. 'When children are stricken with cancer, you fight for their lives and look for reasons,' the couple say. 'We asked all along about the cause. People would say, "This is rare. This is non-hereditary." But we had been so careful as parents.' The Chudas could not imagine what the cause of Colette's cancer might be, so they undertook extensive medical tests to see if there was anything in their own background that might have triggered the disease. The tests came up blank, other than establishing that it was not genetic.

'We began to question whether something in the environment had interfered with Colette's gestational development. We learned . . . that it was possible that something Nancy had ingested or was exposed to in the environment during her pregnancy could have triggered the destructive mechanism that caused Colette's cancer to later develop.' It took four years of harrowing inquiry before a scientific study revealed a link between parental pesticide exposure before or during pregnancy and the age of the child at the time of the diagnosis of Wilms' tumour. Among the many troubling facts the Chudas unearthed was the detail that, in the US at the time, virtually all environmental safety standards were based on research that measured the potential effects of carcinogens on 155-pound [70 kilo] adult males—not on an infant or a foetus.

Their investigations concluded that maternal exposure to pesticides was the most likely cause of the tumour, which was diagnosed when Colette was four years old. This view rested on an extensive study by a Canadian and Brazilian research team, working in Brazil, where the rates of Wilms' tumour are 'among the highest in the world'. Studying 109 cases of the rare cancer, the scientists concluded: 'Consistently elevated risks were seen for farm work involving frequent use of pesticides by both the father and the mother.'[2]

The Chudas said:

> As Colette's parents, we will never forget her bravery. She taught us not to be afraid to die. She proved to us that unconditional love lasts forever. It is this flame that burns deep in our hearts even today. The morning after Colette died, our close friend and neighbor, Marcy Hamilton, came down the hill to our house. She said, 'Colette's favorite color was green. She loved the park. She loved nature. Why don't you start an environmental fund?'

As time went by, the memory of Colette inspired the creation of the Colette Chuda Environmental Fund (CCEF) to support scientific research into the risks to children from environmental toxics. In 1994 this led to the publication *Handle With Care: Children and environmental carcinogens* by the Natural Resources Defense Council which received worldwide distribution.[3] The following year Senator Barbara Boxer proposed far-reaching changes to the US *Toxic Substances Control Act*, which included far greater protection for children and pregnant women. Senator Boxer said:

> Nancy and Jim Chuda, despite their grief, chose to turn their own personal tragedy into something positive. They have labored endlessly to bring to the country's attention the environmental dangers that threaten our children. They want to make sure that what happened to their Colette will not happen to another child. No parent should have to go through what the Chudas went through. If future deaths can be prevented, I know we will all be indebted to the tremendous energy and perseverance of Nancy and Jim Chuda.

That perseverance was to bear fruit still more remarkable. Armed with the latest science about childhood cancers and other conditions, the Chudas set to work to make other parents, caregivers, scientists, environmental lobbyists and the media as aware as they now are of the inherent risks of this chemicalised world into which children are now born—and what can be done to protect them. This led to the launch of *Healthy Child Healthy World*, an informative, sympathetic and practical website that reaches out not only to Americans but to concerned parents and citizens the world over with trustworthy information, advice and support.

Like the pebble that starts the landslide, it heralds not only a change in how society responds to chemical risks, but potentially a broader and deeper change in how we humans as a species now respond to great and complex challenges.

Thinking as a species

At some magical moment in the second trimester of a baby's gestation a marvellous thing happens. The neurons, axons and glia in the embryonic brain begin to connect—and cognition is born. An inanimate mass of cells becomes a sentient being, capable of thought, imagination, memory, rationality, feelings and dreams.

Today, the minds of individual humans are connecting, at lightspeed, around the planet—just as the individual cells establish connections in the foetal brain. We are in the process of forming, if you like, a universal, Earth-sized mind. What the French philosopher and Jesuit Pierre Teilhard de Chardin called the 'noosphere'—or sphere of human thought—is becoming incarnate through global electronic connectivity.[4]

A higher understanding—and potentially a higher intellect—is in genesis: capable of interpreting, and potentially solving, our problems at the supra-human level by applying millions of minds simultaneously to issues, by sharing knowledge freely and by generating faster global consensus on what we need to achieve and the process required to do it.

At the very moment in our social evolution when governments and existing institutions are seen to be failing to tackle the overwhelming issues of overpopulation, resource scarcity, pandemic poisoning, environmental loss and climate change, a new form of human interconnection and self-awareness has emerged that, just possibly, might save us from ourselves.[5]

Even as we now look down on nineteenth-century industry for its filth, cruelty and exploitation of child labour, future generations will come to view our own era as equally filthy and exploitative of children, because we were willing to sacrifice their health to our own consumer habits. To avoid such a judgement of history, citizens the world over need to grasp the scale of the toxic flood now engulfing us all—and that it is within our power to abate it if we so wish. It is within our reach to bring to a close what our descendants will undoubtedly view as a dark age of contamination, preventable sickness, suffering and ignorance.

The means for achieving this new Enlightenment already exist: the internet and social media. By 2017, it is predicted that the internet will reach 3.6 billion citizens of planet Earth (48 per cent of the population)[6] and will probably engage three-quarters by the early 2020s. Through the internet and social media, for the first time in human history, we as a species are sharing thoughts, ideas, problems and solutions around the planet at lightspeed, across almost all the political, religious, cultural, racial, national, economic, social and other barriers which have hindered the sharing of information and understanding in the past.

We are in the process of becoming cognate at the species level—and this is a very exciting and hope-filled moment in our evolution. For the first time, this novel medium of comunication has opened the way for a new human connectedness via shared thought and rapid consensus action on issues which imperil our wellbeing and our future, in real time and at lightspeed. It enables us to collaborate as a single species, for all our individual differences.

True, the internet is rife with trivia, rubbish, ignorance, abuse and misinformation—but it is also laden with science, common sense, idealism, hope, fact, generosity and well-intentioned activity. Like any individual human mind, it contains thoughts which are generous, informed and altruistic—and thoughts which are mean, ignorant and petty. As individuals, most of us try to steer our lives in the direction of the former, and there seems no good reason why globally, collective human thought should not also obey this pattern. Such values are, after all, embedded in most of the religious, ethical, social and governmental institutions we have already formed.

Individuals who value a healthy future for themselves and their children will network online with like-minded people, who also care about children and the future of humanity. Through the power of social media, especially, such people will soon influence many whom such networks do not yet reach or who fear it is beyond their powers as individuals to change our world for the better. In fact, it has never been more possible.

If all the groups of consumers, citizens, parents, victims, farmers, scientists, environmentalists—and others who are concerned about the toxic chemicalisation of the Earth—were to join hands in cyberspace to share information, ideas, advice and mutual support for a cleaner, healthier and safer future, this would be a more influential gathering of people than any political, religious or national movement now or in history. If they were to not only pool ideas, information and thoughts but also to inform purchasing decisions among several billion consumers, this would send strong signals to industry, by rewarding clean products and industrial processes, and penalising polluters and poisoners with loss of market share.

Everyone on Earth is a consumer. We each carry a chemical burden. Most people are, or become, parents. Most of us are potential cancer victims. And all of us are having our genes, minds and bodies subtly distorted by chronic chemical exposure.

By uniting, sharing knowledge, educating one another, and choosing (thus rewarding) clean industries and businesses, we can make a very large difference—one so large, in fact, that most political, religious, governmental and commercial entities will want to be a part of it. Clearly such a global group will only thrive if its objectives are above ideology, prejudice and vested interest, and especially national self-interest. If a clear and present danger to the health of all humans and their offspring into the future cannot unite us to take collective action about this threat to our future, then perhaps we are not the *Homo sapiens sapiens* (meaning 'wise, wise human') we fancy ourselves to be; perhaps we should be looking for a name more suited to an unwise species.[7] There are plenty of global bodies—such as the International Red Cross/Crescent and Oxfam—that prove it is possible for humans from diverse backgrounds and beliefs to come together on matters of mutual concern and responsibility. 'Clean Up the Earth'[8] is simply an undertaking on a larger scale, making full use of the most pervasive means of knowledge-sharing ever devised.

Roles for such a global group might be to:

- Share information universally about toxic chemicals of concern and their health effects as revealed by peer-reviewed scientific evidence.
- Share information on which categories of products are most polluting and should be avoided.
- Share information about which products and processes are cleanest, healthiest and safest and should be encouraged and rewarded, including with higher prices and growing consumer demand.
- Educate parents, consumers and children on how to choose the least-contaminating products.
- Help accelerate the global sharing of scientific knowledge about contaminants among government regulators, international treaty bodies and citizens around the world by reporting new issues and substances of concern.
- Lobby for increased public funding of research into chemo-toxicity and for the mandatory safety testing of *all* new and existing chemical substances and mixtures.
- Lobby for stronger support for preventative healthcare, to supplant the costly and pervasive chemical therapy model.
- Encourage and reward the uptake of 'zero waste' and similar clean philosophies by local communities, industries and governments.

- Establish a '*Forbes* 500'-like list of the world's cleanest companies, to convey the message that health is as important as profit, and to acclaim and honour the best performers as role models and market leaders for others to follow and emulate.
- Provide evidence-based educational material to junior schools worldwide about the risks to children's own health from chemicals (including illegal drugs, alcohol, tobacco, certain processed foods etc.) and thereby help children to educate their parents.
- Create and maintain smartphone 'apps' (applications) that advise consumers which products are safest—and which to avoid—as they shop.

Such a global group or union would *not* engage in consumer bans, physical confrontation, lawsuits or other direct action against industry or science; to do so will only entrench mutual mistrust, delay the move to clean production and drive industry into greater secrecy and into less-regulated parts of the world. It will do best if founded on the principles of cooperation, consensus, openness and equality between society, industry and government.

Such a group need not even be a formal organisation—but rather a networked gathering-together, or a movement of like-minded, well-intentioned bodies and individuals. In fact, if operated on respect, information sharing, common cause and rational persuasion alone, it will be far more influential than as a formal lobby group or NGO. It will also be a lot harder for its critics to discredit it, defuse or shut it down, in the same way industry and political lobbies have already sought to silence and mislead public debate over issues such as global warming, acid rain, tobacco, environmentalism and DDT.[9] Global networks will also apply pressure to local and national governments to decide where they truly stand in their ongoing dilemma over whether to support the public's health or private economic interests.

It need not reach every consumer on earth. A simple majority will do. Or even a groundswell significant enough in number to hold the attention of industries, governments, media and stockmarkets everywhere.

Above all, a 'Clean Up the Earth' network ought not be an official body or anything that smacks of 'world government'. If the people of the world wish not to be poisoned, they can express their wish through knowledge sharing and their own behaviour in the market without need of authoritarian structures or dogma. 'Clean Up the Earth' may operate more effectively as a loose alliance dedicated to a single cause. It could even become the first genuine exercise in global grassroots democracy, founded upon shared values and interest—of a kind no existing government structure can yet deliver.

If as a society we truly value our health and that of our children, then the issue of toxic exposure is an immediate concern, wherever or whoever we are. As a concept, the risk of being poisoned is not difficult for the individual to comprehend. It therefore has a better shot at building a planet-wide consensus than almost any other issue—enabling us to take a united step into a new human future of self-governance by mutual agreement. Cynics will certainly dismiss this as naive idealism, but it will seem like naivety only to those who do not yet fully appreciate the significance of the species-level changes now taking place in humans' ability to conduct a planet-wide discourse—or those who cling to the atrophying, historical structures of power (such as the nation state) which are so plainly failing to deal with current and impending threats to human health, welfare and existence at the species level.

Speaking out

In recent years there has been a proliferation of groups around the world representing all sorts of interests around chemicalisation: concerned parents, anxious consumers, sufferers of various diseases, industrial workers, environmentalists, organic farmers, food fanatics, religious bodies, healthcare professionals, green industries and some scientists. These groups have enjoyed significant success in sensitising educated citizens to the risks posed by chemicalisation and in initiating a change in purchasing habits in areas such as food—but for all their outspoken critiques, they have had less success in persuading governments to curb the flood of new chemicals onto the market or to withdraw many old and now highly suspect substances. The achievement of the US Breast Cancer Fund in persuading the US government to adopt a preventative approach to the disease appears a promising step—but it remains to be seen whether it has much impact on the global, or even the American, chemical burden. In Europe the banning of several hundred agricultural pesticides has led to an equally promising reduction in the presence of substances such as DDT and dioxins in the food chain—offset by an increase in new chemicals of unknown toxicity. It is now time for a citizens' movement to go to the next level.

The counter lobby

The critics of chemicalisation have been the subject of serious counter-action by some chemical and other pro-industry lobbies. Typifying this, a blog on the website of the journal *Scientific American* mockingly deplores the 'growing epidemic of antiscientific, fact-free chemophobia that abounds on the Internet. In fact most of it is not even chemophobia, it's just plain ignorance of basic science. But as far as the sheer amount of chemistry-related nonsense floating around goes, all this worthy debunking is no more than a drop in the ocean. Those who don't understand science and chemistry are going to keep foisting the same falsehoods on us ad nauseum [*sic*] . . .'[10] And while not all chemical bodies are as blatant as this in publicly belittling their critics as stupid, uneducated and ignorant for wanting to avoid being poisoned, it must be said that such industry views have so far weighed more with governments and regulators than the calls of citizens' groups to lower the societal chemical burden and its associated rates of disease and death.

Typically, both industry spokespeople and the critics of public chemical concerns rely on the dose-response argument, which dates back to the renaissance doctor Paracelsus (1493–1541). This states that a person's response to exposure to a certain chemical depends on two factors: the quantity of the substance and the length of their exposure time. This principle is often used to argue that very tiny, infrequent doses of toxins do no harm—it is often then taken a step further to dismiss the risk from continual exposure to a host of chemicals in tiny amounts.

This argument has many flaws. Firstly, the degree and type of harm are not always easy to diagnose (and may pass undetected by current methods until it is too late), or else the damage may require a long period of chronic exposure to manifest and to be linked to a specific trigger. One of the lessons from the endocrine disruptors is that, paradoxically, they can sometimes exhibit a larger response at a lower dose. Secondly, the human dose-response, and the child dose-response, are simply unknown for most chemicals. Thirdly, the compounding effect of chemicals in mixtures must be taken into account, as this is nearly always how chemicals are encountered by people. And fourthly, we cannot ignore the recent research in the rapidly advancing field of epigenetics, which has shown that tiny amounts of certain substances can have important effects on gene behaviour, sometimes spanning generations.

The dose-response argument is sometimes coupled with the specious claim that 'human life expectancy is increasing so chemicals can't be doing us any lasting harm'. This ignores the possibility that currently reported life expectancy figures, being based on retrospective data, may change in future (and indeed, many fear life expectancy will decrease due to the spread of obesity and diabetes alone), or that the benefits from improvements such as clean water and air or better medical treatment may mask deterioration in other factors. Whatever the arguments about dose-response and life expectancy, they constitute neither a rational nor a moral basis for ignoring humanity's increasing chemical exposure, the resulting human death toll, or the public's growing concern about it.

Sadly, it appears that 'Minamata disease'—here referring to the stalemate confrontation between the corporate world and wider community—risks becoming pandemic, thereby paralysing the ability of humans to act in our own best interests. As shown in numerous chemical poisoning episodes, the louder and angrier the public protests become, the more resolute, secretive and dug-in the source industries become—and the greater their reluctance to listen or negotiate. As industry typically has economic resources to throw into legal and political fights that vastly outweigh those of its critics, this unproductive standoff can go on a long time without benefit to anyone. This underlines that traditional forms of consumer and green activism, while useful in raising public awareness, might prove counter-productive in reducing the overall chemical burden, particularly if they generate a stand-off where nobody wins and nothing much changes. In a situation in which everyone carries a body burden of toxins, it is in our common interest to find agreed solutions.

Worldwide growth in the internet and social media has achieved remarkable progress in raising awareness and empowering informed purchasing among some consumers, notably the educated, the well-off and the young and aspirational. Informed purchasing is now beginning to exert real market pressure on retailers and manufacturers to withdraw toxic products from sale. In some cases, this has successfully resulted in companies taking the hint offered by the home market, and making a virtue of necessity by marketing cleaner, greener goods and services. In other cases, producers of toxic products have simply relocated offshore or moved into other markets while continuing to pollute the planet.

In an impressive example of consumer power being intensified by social networking, in 2013 a supergroup of almost fifty consumer and health lobbies in the US took aim at the country's ten largest retailers—including Wal-Mart, Target and Costco—in a social media campaign designed to force the retailers to withdraw hundreds, possibly thousands, of products containing 100 toxic substances: formaldehyde, parabens, phthalates, BPA, flame retardants and others.[11] The lobbyists contended that the US government had failed to sufficiently protect its citizens by not requiring the removal of products linked to cancer and other health risks and, in any case, big retailers had more power than anyone to change markets and products by sending the right signals upstream to manufacturers and producers.

For retailers, this sort of pressure is hard to resist, especially if they find themselves targeted by groups with high public profile, support and sympathy such as the Breast Cancer Fund. Evidence that this new form of 'retail therapy' can work exists in almost every supermarket in the developed world in the form of its organic and health food display. Ordinarily, supermarkets don't like handling organic produce—they say it lacks the uniformity, regular supply, blemish-free appearance and the cheap price of the chemicalised product. But when it comes to trends in consumer demand, and winning or losing market share, they are all ears. Surveys show that up to 95 per cent of consumers believe organic food is healthier and safer to eat[12] and global data suggest that farmers are responding by increasing the world area under organic production methods year by year.

There is an unmistakeable trend among middle-income consumers, especially in Europe but also in North America and Oceania and the more affluent cities of Asia, to buy more organic produce and shop at farmers' markets or even buy from organic producers online for home delivery. Global organic food sales in developed countries reached $55 billion in 2009, up 10 per cent in a single year. The countries with the largest markets are the United States, Germany, and France while highest per capita consumption is in Denmark, Switzerland and Austria. This is a market signal that no self-respecting food retailer can ignore and some are even marketing their own organic home brand products in response, as well as sourcing new lines from countries still uncontaminated by Western industrial farming methods.

Further along the supermarket aisle is fresh evidence of consumer preferences driving retail and farming behaviour: the egg display, where free range, deep litter and organic products are rapidly pushing those of traditional battery poultry production off the shelf, even though the latter are cheaper. Though it still represents a small proportion of all the food and goods sold by supermarkets, the natural food and product movement represents a potent demonstration of the power of the motivated consumer to change the product, and even the production process.

The concern driving the natural food movement is rooted in the educated, middle-to-high income classes of developed and newly-industrialising countries. Numerically they are small, but their buying power is big and is gaining both political and market influence. Like many middle-class revolutions, this one may ultimately decide the outcome. Potentially it represents the bow-wave of a global supertanker of opinion that could change not only markets for food, but also for almost every other product or service.

Citizen action

Experiences such as those of Nancy and James Chuda, Lisa Leake or Sue and Howard Dengate and other ordinary, loving parents are lending fresh impetus to the cause of raising worldwide consumer awareness about the need to reduce the toxic burden in our lives. This new level of consumer awareness is closely coupled with the rise in 'ethical consumerism', in which informed consumers purposefully avoid products produced in ways that conflict with their moral values. Major examples in recent times have been the consumer campaigns against sporting goods manufacturer Nike[13] and the backlash against the world fashion industry following several Bangladesh tragedies in which hundreds of poor garment workers lost their lives in fires and building collapses.[14] One of the most important findings from these sad events was a survey by Oxfam which showed that 70 per cent of consumers would be willing to pay more for garments if they could be assured they were produced ethically.[15] 'Consumer sentiment is becoming much more finely attuned to how things are produced,' says Oxfam Australia CEO Helen Szoke. While there is still scepticism over whether most people are genuinely ready to put their money where their mouth is and shop ethically even if they pay more, the question of whether they are willing to be poisoned personally, and their children too, falls much closer to home. Given an informed choice, it is probable most people and almost all parents, will prefer to pay a bit more to reduce their toxic exposure.

In addition to the growing resolution displayed by concerned parents and citizens, large international lobbies and pressure groups are also forming and growing, which betoken the emergence of a worldwide consensus on reducing the global chemical burden. Prominent groups involved in this activity include:

- AVAAZ, a group established with the specific object of 'bringing people-powered politics to decision-making worldwide'. The name means 'voice' in several languages, and the group was founded in 2007 to 'organize citizens of all nations to close the gap between the world we have and the world most people everywhere want'. It claims a membership of around thirty million. Eschewing issue-based lobbying, AVAAZ works across a range of issues of public concern using its community 'to act like a megaphone to call attention to new issues'. Unfortunately a poll of its members indicated food and health were among their lowest priorities (it did not mention chemicals specifically at all) although there was some support for a global climate treaty, and for the European campaign to ban neonic pesticides.[16]
- SumOfUs is 'a movement of consumers, workers and shareholders speaking with one voice to counterbalance the growing power of large corporations'. It campaigns against 'what happens when powerful corporations get their way' listing 'poisons pouring into our air and water', and was also involved in the anti-neonic campaign.[17]
- GetUp! is an Australian web-based movement claiming 640,000 followers and billing itself as 'an independent movement to build a progressive Australia and bring participation back into our democracy'. It mainly campaigns on current political and public issues, which include global warming and fracking for natural gas.[18]

- The Safer Chemicals, Healthy Families coalition is a US-based group claiming to represent 'more than eleven million individuals and includes parents, health professionals, advocates for people with learning and developmental disabilities, reproductive health advocates, environmentalists and businesses'.[19] Interestingly, it combines scientists, healthcare providers and citizens and has a specific focus on the chemical exposure issue.
- With more than four decades of high-profile campaigning under its belt, Greenpeace is one of the world's most widely recognised and influential environmental groups. Based in the Netherlands, it has offices all round the world and claims 2.8 million supporters. It has a specific campaign against 'Toxic chemicals in our environment (that) threaten our rivers and lakes, our air, land, and oceans, and ultimately ourselves and our future',[20] as well as campaigns for clean water and greener electronics.
- Friends of the Earth International (FoEI) is a global federation of seventy-four national environmental organisations, founded in 1969 and claiming more than two million members worldwide. 'We challenge the current model of economic and corporate globalization, and promote solutions that will help to create environmentally sustainable and socially just societies,' it says.[21]
- Founded in 1982, the Pesticide Action Network (PAN) has nodes in five continents and networks 600 citizen and farmer organisations across ninety countries who are concerned about pesticides.[22] It also has a useful database of 6500 agricultural and food chemicals.[23]

- Circle of Blue is an independent non-profit journalism service reporting on water issues worldwide, including chemical pollution.[24]
- The International River Foundation works in partnerships around the world to fund and promote the sustainable restoration and management of river basins including ways to reduce their contaminant load. It awards prizes for outstanding achievement in river clean-up and shares knowledge among river communities on best practice.[25]
- The WWF (Worldwide Fund for Nature) was formed in 1961 from several existing conservation bodies to promote and fund nature conservation. It currently works in around 100 countries and claims five million members worldwide. Its work includes supporting cleaner ways to produce food and other goods, reduce contamination by industry and energy generation and deal with waste. It has a philosophy of forming strong partnerships with businesses, governments and local communities to drive change.[26]

This list is by no means exhaustive, and is intended rather to illustrate the range and diversity of international organisations with an interest—existing or potential—in preventing and reducing the poisoning of our planet and its people. Together these organisations claim more than fifty million members globally and, if they ever act consistently and in concert on a single issue, will constitute a formidable expression of people power. To the list above must be added countless new and growing groups such as Nancy and Jim Chuda's *Healthy Child Healthy World*, Lisa Leake's *Real Food* and Sue and Howard Dengate's *Food Intolerance Network*

(FedUp!), not to mention thousands of similar groups, small and large, on Facebook, Twitter and other social media.

While the expression of global people power via the internet is still in its infancy, there is little doubt as to its trajectory—and that it will in time become a force that governments, corporates and professions such as chemistry will find progressively difficult to ignore. In less time than it may appear, a planetary consensus is being formed around several issues and a new, hopefully more evolved, form of democracy is being born. On matters such as chemicals and climate, humanity is learning to think as a species, for its collective wellbeing.

A matter of science

Over several decades of intensive research into facets of the issue, scientists have managed to demonstrate that the impact of chemical contamination has spread from a local issue to a global one. However, while international regulatory bodies have been formed, there has as yet been scant international scientific effort to study Earth system contamination in detail on anything like the scale devoted to climate change. This is all about to change, with the formation in Australia in 2013 of the Global Contamination Initiative (GCI). This is the brainchild of Professor Ravi Naidu, an eminent Australian contamination scientist and a researcher recognised internationally for his leadership in the clean-up field. He says:

> Contamination by the chemical products and by-products of human activity is one of the most pervasive and far-reaching of our impacts on the Earth and on our own health and wellbeing.

> Traces of anthropogenic contamination are now to be found from the stratosphere to the deep oceans, from pole to pole, in many forms of wildlife, in all modern societies, the food chain and in most individuals, including newborns . . . Chemical contamination is of equivalent significance with climate change, nutrient pollution, biodiversity loss or any of the other major impacts of human population growth and development upon the Earth's biosphere, human health and wellbeing.

The GCI is a global scientific initiative launched at the international CleanUp conference in 2013, with support of scientists from many countries, to define, quantify, set limits to, help clean up and devise new ways to curb the growing chemical assault on human health and the biosphere. 'We envisage this as an international alliance of leading scientific, government, industry and community organisations and individuals dedicated to making ours a cleaner, healthier and safer world,' Professor Naidu explains. He continues:

> The initiative seeks not only to define the extent of contamination at international scales, but also to develop and share cost-effective, workable solutions which can be readily adopted by industry, governments and the community. These include further developing and disseminating the concept of 'green production'—the production of goods and services without accompanying risk of contamination. The Global Contamination Initiative (GCI) is a worldwide knowledge network, performing new scientific research, aggregating existing knowledge, developing novel assessment and clean-up technologies, advising governments and industry on ways to improve existing regulation or industry practices, training

> high-level experts and sharing information about ways to reduce anthropogenic contamination in all facets of human society and the natural environment.

Among the GCI's likely undertakings are developing a 'stocks and flows' model of global contamination, efforts to establish safe boundaries for key pollutants, identification of urgent issues and areas for action, investigation of the combined effects of contamination on human health and the environment and the international sharing of solutions. The Initiative will work in close partnership with other major scientific programs (such as the seven major infant toxicity studies referred to in Chapter 3) and with leading universities and corporations.

'There is a widespread lack of awareness among governments and societies about the current scale, pervasiveness and risk to billions of people from contamination in the Earth system,' Professor Naidu says. 'Indeed so complex is the threat that we do not yet fully understand it or how best to curb it. There is still a grave lack of international institutions committed to tackling it—and a lack of the science needed to underpin their work. There is also a lack of a real appreciation of the many benefits and economic, social and environmental gains to be obtained from cleaning up our world. The challenge is urgent—and we need to start now, working at global as well as local level.'[27]

A matter of rights

Every person in the world has a right to life, liberty, personal security, to marriage and family, to equality, to work, to education, to the law, to elect a government, to freedom of opinion or belief, to asylum. These are just some of the rights available to each of us under the thirty articles of the *Universal Declaration on Human Rights*,[28] though it is true that many governments still fall short of these prudent and fair requirements when measured by how they treat their own citizens and others. Nevertheless, such human rights represent an important aspiration for all the world's inhabitants in the twenty-first century, as well as a yardstick by which governments and regimes may be measured and judged.

It is therefore more than a little disturbing to find there is no human right not to be poisoned.

A child born today may enjoy many of the rights listed under the Universal Declaration—but not the right to a full intelligence, to undamaged genes, to a life free of cancer, mental or reproductive dysfunction or other attributes increasingly linked by science to chronic lifelong chemical exposure. The lack of such a right, in the presence of a right to leisure, to social security or to cultural participation bespeaks a remarkable blind spot in the contemporary conscience, deriving either from an acute lack of awareness of the scale of the problem or from a general desire to hide from ourselves knowledge which is unpleasant, distasteful or disturbing, especially if it means having to change the way we produce material goods, food, energy and how we go about things in our daily lives.

In Article 5, for example, the Universal Declaration proclaims that every human has a right not to be tortured. Although this is a right which, presumably, applies to only a small percentage of the world population at any one time, there is no articulated right for people to be safe from the flood of toxins or suspected toxins that now inundates virtually the entire human race, cradle to grave. There is no right to be safe from an assault that we know is killing millions and impairing the health of tens of millions more—often in ways that might well be deemed torture if you caught someone deliberately inflicting them.

Likewise everyone has a right to education—but apparently not a right to the inborn intelligence that would enable us to take full advantage of it.

We all have a right to security of person—except when we inhale, ingest or absorb something toxic from the throbbing industrial machine that now encircles the globe.

We have a right to equality before the law, except when it comes to challenging the producers of toxins, who have so often used the law as a shield against their critics and against the necessity to change their products or processes.

We have the right to marry and have a family—but not one whose health, wellbeing and future are protected from the contaminants, both legal and illegal, now circulating in the Earth system.

We have a right to equal access to the public service of our country, except in cases where the public service, often acting under political pressure, takes the side of a contaminator against the contaminated or fails to investigate thoroughly and objectively the risks and concerns its citizens may place before it.

We each have a right to a job—but no clear right not to be poisoned while doing it.

We each have a right to 'a standard of living adequate for the health and wellbeing of himself and of his family', but not a right to a standard of living as free from toxins as almost all previous generations of humans have enjoyed.

The existence of human rights does not prevent these rights from being abused in a great many cases—but it does establish an acceptable worldwide standard of behaviour for humanity in general, and it goes a long way towards encouraging their adoption by countries, communities, corporations and individuals, spurred by the threat of exposure of those who breach them.

One of the principal difficulties in attempting to enshrine a right not to be poisoned is that there is at present not a single government or regime on Earth that could truthfully claim to uphold it. Even the most scrupulous, strict and efficient administrations have no solution to the twin problems of Earth system contamination and toxic mixtures—and, as the UN Environment Programme has pointed out, administrations are still broadly in ignorance of the toxicity of most of the chemicals in regular use—or even now being introduced—in their own jurisdictions. For this reason any proposal to introduce such a right may face widespread opposition from nation states, backed up by industries and interests which are reluctant to change.

But that does not mean that concerned citizens around the world should not seek the instatement of such a right, or seek to arouse others to an awareness of the necessity. Rights, as we know from history, are not to be had for the asking—they are to be argued and fought for, often against bitter and entrenched opposition, usually over many years and sometimes over generations. They represent a standard to which all people, nations and corporations may aspire.

If we do not have a 'right not to be poisoned', there will probably never again be a day when we are not.

Preventing the Poisoned Planet

If we wish to lower the toxic burden for ourselves, our children and all life on Earth, we should together take the following actions:

1. Form a global network of people and institutions concerned about cleaning up the Earth to spread awareness, motivate industry to adopt clean production systems and help citizens to become 'clean consumers'.
2. Campaign for a universal human right not to be poisoned.
3. Press for the replacement of all coal, oil and other fossil fuels with clean energy and with clean feedstocks for industry, aided by informed consumer preferences, as soon as possible.
4. Progressively eliminate toxic substances in the food chain and wider environment through informed consumer preference and enhanced regulation.
5. Press for a priority policy of disease prevention in medicine, over chemical cure. Educate health workers to recognise, diagnose, report and prevent diseases resulting from chronic chemical exposure so as to inform the public debate. Train them to rely less on chemical medicine.
6. Train all young chemists, scientists and engineers in their social and ethical responsibility to 'first, do no harm'.
7. Educate our children to choose wisely among products and services, based on their personal and universal health impact. Empower them to educate us.

8. Empower industry to make profits ethically, by producing clean products that do no harm.
9. Reward industries which adopt approaches such as green chemistry, product stewardship and zero waste with our custom and support.
10. Press for the universal toxicity testing of all suspected and new industrial substances and major waste streams.

There are plenty of other things we could do, but ten seem sufficient for the purpose of this short book, which is to herald a clear and present danger to ourselves and to future generations—and also to set forth the inspirational possibility of our overcoming it, together, in ways that lead to better health, greater prosperity and to a cleaner, more sustainable and fairer world.

The poisoned planet is a fate we can avoid if we, the citizens of Earth, so will it.

POSTSCRIPT
A CAUTIONARY TALE FROM DEEP TIME

It was a fine Tuesday morning about 2400 million years ago. A faint but increasingly vigorous young sun shed its silvery radiance through gaps in the broiling clouds and layers of mist that still enshrouded the infant Earth, bestowing light and heat to warm the sultry waters of a virulently coloured sea. Within those waters vast mats and clumps of livid algae and purple microbes basked in the new sunlight, soaking up its energy and quietly using it to make and digest useful food. Every so often one of them would emit a tiny, satisfied belch. Invisibly, tiny molecules of oxygen entered the water and, finding nothing much to cling to, joined their companions to form a bubble which rose slowly but resolutely to break the oily surface of a stagnant world ocean. As far as the eye could see—had there been any eye to see it—the watery world was all bubbles, thrusting through the primordial sludge of early life, bursting into the atmosphere like the expiring fizz from a bottle of lukewarm soda water.

In many ways it was a beautiful Eden: the sky had a greenish tinge, the ocean formed a rainbow swirl of reds, blues, greens and yellows reflecting the teeming life and the patterns of light that played upon it as pulses of sunlight rippled through the racing cloudscapes. The continents were dusty reds and greys, devoid of all vegetation and stripped to the bone by the sandblast of cyclonic winds and the torrential rains that levelled mountain ranges higher than the Himalaya in the twinkling of a geological eye. Life was somnolent but rich for the countless billions of stromatolites, blue-green algae and other microscopic organisms which burgeoned in muddy pools, tepid lakes and tidal wetlands, cloaked the surface of bays and inlets, slurped the nutritious richness of minerals from volcanic springs and seabed smokers.

For more sophisticated life forms, such as ourselves, it would have been a lethal Garden of Eden: simply drawing a breath from the atmosphere rich with nitrogen, carbon dioxide and water vapour but almost devoid of precious oxygen, would kill you in no time. Even if you lived, you'd fried by the blistering ultraviolet rays from a young sun that sparkled through the spaces between the clouds.

Stromatolites and blue-green algae were the descendants of Luca, the 'last common ancestor', a creature dwelling so far back in the misty past as to be almost a scientific legend. Luca itself evolved on—or came to—the Earth about 4000 million years ago, as soon as the first torrential rains had filled the ocean basins, formed lakes and rivers and created a suitably damp home. Luca was a lean and hungry beast, largely given to devouring its own kind for the organic carbon it needed to survive.

Then, about 500 million years in, a miracle occurred. The youthful sun was just starting to break through the dense, steamy cloud base and Luca took advantage of this potent new energy source, developing the vital process on which most life now depends: photosynthesis, the ability to transform carbon dioxide and water into food, using sunlight. Like modern industrial systems, photosynthetic life forms—otherwise known as plants—take in what they need and excrete the rest as waste. In their case, however, the 'waste' is oxygen.

For a billion years or so it didn't really matter how much oxygen the ancient photosynthetic organisms put out: it was quickly absorbed by vast quantities of minerals lying around from the early formation of the Earth. This led, among other things, to the gigantic limestone formations, and the iron ore deposits which we mine today to make steel. For another billion years or so these minerals absorbed all the free oxygen the early micro-organisms could churn out—and things remained pretty much unchanging and humdrum, so far as life was concerned.

Then, about 2400 million years ago, like a garbage tip that has taken all the rubbish it can hold, the minerals reached saturation and, slowly, molecule by molecule, the amount of oxygen in the atmosphere began to climb. For a time this didn't matter much to early life: it continued much as it always had, soaking up CO^2 and excreting O^2, warmed and driven by the sun. Indeed, there was even an advantage, as the accumulating oxygen formed ozone which served as a shield against the deadly ultraviolet rays—a sort of planetary parasol built from the excretions of living organisms.

POISONED PLANET

After a while, however, like one of those old mediaeval cities where people threw their filth into the public streets, oxygen levels began to climb and the planet gradually grew foul. The early life forms began, literally, to suffocate in their own mess and to die in droves. From being of little significance, the oxygen had suddenly emerged as a deadly poison—catastrophically so.

At this point the whole of life on earth underwent what scientists have called the 'oxygen holocaust', a period of universal global toxicity that led to mass extinction. The worst die-off ever in the history of Earth. Only a handful of organisms survived, notably those that adapted to breathing oxygen. Some of them went on to become us.

And we, unthinkingly, went on to pollute the world anew. To create a man-made chemical holocaust.

The moral of this story?

If you foul your planet, it may kill you.

ACKNOWLEDGEMENTS

I wish to express particular thanks to . . .

Professor Ravi Naidu
Professor Ming Hung Wong
Professor Jack Ng
Professor Megharaj Mallavarapu
Dr James Siow
Richard Walsh
Elizabeth Weiss
Siobhan Cantrill and the team at Allen & Unwin
John Manger
Tim Curnow

NOTES

Chapter 1

1 W. Eugene Smith and Aileen M. Smith, *Minamata: The story of the poisoning of a city, and of the people who choose to carry the burden of courage*, Holt, Reinhart and Winston, New York, 1975.
2 Ibid.
3 *Your Dictionary*, <http://biography.yourdictionary.com/w-eugene-smith>.
4 US Environment Protection Agency (US EPA), *Toxic Substances Control Act (TSCA) Chemical Inventory,* December 2012, <www.epa.gov/oppt/existingchemicals/pubs/tscainventory/index.html>.
5 Agency for Toxic Substances and Disease Registry (ATSDR), *Chemicals, Cancer and You*, 2010, <www.atsdr.cdc.gov/emes/public/docs/Chemicals,%20Cancer,%20and%20You%20FS.pdf>.
6 A. Kortenkamp, M. Faust and T. Backhaus, *State of the Art Report on Mixture Toxicity*, European Union, Geneva, 2009, Executive Summary, p. 11.
7 Department of Health, *Australian Inventory of Chemical Substances* (AICS), 2013, <www.nicnas.gov.au/regulation-and-compliance/aics>.
8 UN Environment Programme (UNEP), *Global Chemicals Outlook: Towards sound management of chemicals*, 2012, <www.unep.org/pdf/GCO_Synthesis%20Report_CBDTIE_UNEP_September5_2012.pdf>.
9 Ibid.
10 ATSDR, *Chemicals, Cancer and You*, p. 1.
11 UNEP, *Global Chemicals Outlook,* p. 10.
12 Ibid., p. 19.
13 X.T. Tran, *Consequences of Chemical Warfare in Vietnam*, March 2006; X.T. Tran, *Agent Orange: Diseases associated with Agent Orange exposure*, 25 March 2010, both

US Department of Veterans Affairs Office of Public Health and Environmental Hazards.

14 Assumes tripling in global chemical output (UNEP) and global population of 9.6 billion (UN Population Program).

15 J. Rockstrom et al., 'A safe operating space for humanity', *Nature*, 461, 2009, <www.stockholmresilience.org/download/18.1fe8f33123572b59ab800012568/pb_longversion_170909.pdf>.

16 R.J. Diaz and R. Rosenberg, 'Spreading Dead Zones and Consequences for Marine Ecosystems', *Science*, 321(5891), 2008, pp. 926–9.

17 B.H. Wilkinson and B.J. McElroy, 'The impact of humans on continental erosion and sedimentation', *Geological Society of America Bulletin*, July 2006, <http://gsabulletin.gsapubs.org/content/119/1-2/140>.

18 MBendi Information Services, 'World Mining—Overview', <www.mbendi.com/indy/ming/p0005.htm>.

19 W. Liu, 'The advancement and developing of red mud utilization in China', ICSOBA, 2013, <www.redmud.org>.

20 Earthworks and Mining Watch Canada, *Troubled Waters: How mine waste dumping is poisoning our oceans, rivers, and lakes*, February 2012, <www.earthworksaction.org/files/publications/Troubled-Waters_FINAL.pdf>.

21 'Pollutants from coal-based electricity generation kill 170 000 people annually', *Greenblog*, 14 June 2008, <www.green-blog.org/2008/06/14/pollutants-from-coal-based-electricity-generation-kill-170000-people-annually/>.

22 M. Bittman, 'Giving up tuna? Breathing is next', *New York Times*, 11 June 2013.

23 N. Pirrone et al., 'Global mercury emissions to the atmosphere from anthropogenic and natural sources', *Atmospheric Chemistry and Physics*, 10, 2010, pp. 5951–69, <www.atmos-chem-phys.net/10/5951/2010/acp-10-5951-2010.pdf>.

24 D.K.Y. Tan and J.S. Amthor, 'Bioenergy', in Z. Dubinsky (ed.), *Photosynthesis*, Intech, Rijeka, Croatia, 2013.

25 Tyndall Centre for Climate Change Research, University of East Anglia, 'Record high for global carbon emissions', 2012, <www.tyndall.ac.uk/communication/news-archive/2012/record-high-global-carbon-emissions>.

26 Organisation for the Prohibition of Chemical Weapons (OPCW), 'Demilitarisation: Latest facts and figures', 2013, <www.opcw.org/our-work/demilitarisation/>.

27 One of these is Syria, whose stockpiles may finally come under international control as a result of the UN agreement formed during the country's civil war.

28 World Nuclear Association, 'Supply of uranium', <www.world-nuclear.org/info/Nuclear-Fuel-Cycle/Uranium-Resources/Supply-of-Uranium>.

29 International Atomic Energy Agency, *Estimation of Global Inventories of Radioactive Waste and Other Radioactive Materials*, IAEA 2008 and World Nuclear Association, 2013.

30 For a breakdown of reported hazardous waste, see <www.grida.no/graphicslib/detail/global-hazardous-waste-generation-by-type-as-reported-by-the-parties-to-the-basel-convention-for-the-years-1993-2000_1031>.

31 Burning 1 tonne of coal liberates 2.86 tonnes of CO_2, according to the US Energy Information Administration, <www.eia.gov>.

Chapter 2

1 Northeast Fisheries Science Centre, 'Persistent man-made chemical pollutants found in deep-sea octopods and squids', June 2008, <www.nefsc.noaa.gov/press_release/2008/SciSpot/ss0810>.

2 R. Naidu and M.H. Wong, *Remediation of Contaminated Sites*, CRC CARE, Salisbury, SA, 2013.

3 G. Prokop, M. Schamann and I. Edelgard, *Management of Contaminated Sites in Western Europe*, European Environment Agency, Copenhagen, June 2000.

4 W. Klein, 'Mobility of environmental chemicals, ecotoxicology and climate', in P. Bordieu, J.A. Haines, W. Klein and C.R. Krishna Murti (eds), *Ecotoxicology and Climate*, SCOPE 38, John Wiley & Sons, Chichester, 1989, pp. 65–78.

5 US EPA, 'The science of ozone layer depletion', 2011, <www.epa.gov/ozone/science/index.html>.

6 World Meteorological Organisation (WMO), 'Scientific assessment of ozone depletion', 2010, <www.esrl.noaa.gov/csd/assessments/ozone>.

7 UNEP Ozone Secretariat, 'The 2010 assessment of the Scientific Assessment Panel', 2010, <http://ozone.unep.org/Assessment_Panels/SAP/Scientific_Assessment_2010/index.shtml>.

8 World Health Organization (WHO), 'Ultraviolet radiation and human health', 2013, <www.who.int/mediacentre/factsheets/fs305/en>.

9 S. Mosley, 'Environmental history of air pollution and protection', in *Encyclopedia of Life Support Systems*, EOLSS, Oxford, 2012.

10 H. Akimoto, 'Global air quality and pollution', *Science*, 302, 2003, p. 1716.

11 UNEP Centre for Clouds, Chemistry and Climate, *The Asian Brown Cloud*, 2002.

12 'The east is grey', *The Economist*, 10 August 2013, <www.economist.com/news/briefing/21583245-china-worlds-worst-polluter-largest-investor-green-energy-its-rise-will-have>.

13 WHO, 'Mortality and burden of disease from outdoor air pollution', 2012, <www.who.int/gho/phe/outdoor_air_pollution/burden_text/en>.

14 WHO, 'Air pollution', 2013, <www.who.int/ceh/risks/cehair/en>.

15 The Ecology Center, 'New guide to toxic chemicals in cars helps consumers avoid "new car smell" as major source of indoor air pollution', 2012, <www.healthystuff.org/press.releases.php>.

16 R. Dietz et al., 'Three decades (1983–2010) of contaminant trends in East Greenland polar bears (*Ursus maritimus*)', *Environment International*, 59, 2012, pp. 485–93.

17 F. Provieri and N. Pirrone, *Mercury Pollution in the Arctic and Antarctic Regions: Dynamics of mercury pollution on regional and global scales*, Springer, New York, 2005.

18 Australian Antarctic Division, 'Pollution and waste', 2012, <www.antarctica.gov.au/environment/pollution-and-waste>.

19 R. Fuoco et al., *Persistent Organic Pollutants in the Antarctic Environment*, Scientific Committee on Antarctic Research, Cambridge, 2009, <www.scar.org/publications/occasionals/POPs_in_Antarctica.pdf>.

20 UNEP, 'What are POPs?', Stockholm Convention, <http://chm.pops.int/Convention/ThePOPs/tabid/673/Default.aspx>.

21 S. Clarke, 'Antarctic melting 10 times faster than 600 years ago', *ABC News*, 2013, <www.abc.net.au/news/2013-04-15/antarctic-melting-ten-times-faster-than-600-years-ago/4628404>.
22 University of Bristol, 'Future sea level rise from melting ice sheets may be substantially greater than IPCC estimates', *Science Daily*, 6 January 2013, <www.sciencedaily.com/releases/2013/01/130106145743.htm>.
23 B. Yeo and S. Langley-Turnbaugh, 'Trace element deposition on Mt Everest', *Soil Survey Horizons*, 51(4), 2010, pp. 95–101.
24 H. Auman et al., 'PCBS, DDE, DDT, and TCDD—EQ in two species of albatross on Sand Island, Midway Atoll, North Pacific Ocean', *Environmental Toxicology and Chemistry*, 16(3), 1997, pp. 498–504.
25 D.C.G. Muir et al., 'Toxaphene and other persistent organochlorine pesticides in three species of albatrosses from the North and South Pacific Ocean', *Environmental Toxicology and Chemistry*, 21(2), 2002, pp. 413–23.
26 P.S. Ross et al., 'Harbor seals (*Phoca vitulina*) in British Columbia, Canada, and Washington State, USA, reveal a combination of local and global polychlorinated biphenyl, dioxin, and furan signals', *Environmental Toxicology and Chemistry*, 23(1), 2004, pp. 157–65.
27 Seafish, 'Contaminants', 2012, <www.seafish.org/industry-support/legislation/contaminants>.
28 SeaWeb, 'Contaminated sediments', *Ocean Issue Briefs*, 2013, <www.seaweb.org/resources/briefings/sediment.php>.
29 SOE Norway, 'Polluted marine sediments', 2009, <www.environment.no/Topics/Marine-areas/Hazardous-chemicals-in-coastal-waters/Polluted-marine-sediments>.
30 O. Malm, 'Gold mining as a source of mercury exposure in the Brazilian Amazon', *Environmental Research*, 77(2), 1998, pp. 73–8, <www.sciencedirect.com/science/article/pii/S0013935198938282>.
31 S.H. Verhovek, 'After 10 years, the trauma of Love Canal continues', *New York Times*, 5 August 1988, <www.nytimes.com/1988/08/05/nyregion/after-10-years-the-trauma-of-love-canal-continues.html>.
32 Environmental Working Group (EWG), 'Cancer-causing chemical found in 89 percent of cities sampled', 2010, <www.ewg.org/chromium6-in-tap-water>, diagram at <www.pbs.org/newshour/multimedia/chromium-cities>.
33 M. Sviridova et al., 'UK Groundwater Pollution', n.d., <http://sitemaker.umich.edu/section2group1/solutions_and_world_implications>.
34 B. van Wyck, 'The groundwater of 90% of Chinese cities is polluted', *Danwei*, 18 February 2013, <www.danwei.com/the-groundwater-of-90-of-chinese-cities-is-polluted>.
35 'Groundwater in 33% of India undrinkable', *Times of India*, 12 March 2010, <articles.timesofindia.indiatimes.com/2010-03-12/india/28132254_1_fluoride-drinking-water-salinity>.
36 Pacific Institute, 'World water quality facts and statistics', 2010, <www.pacinst.org/reports/water_quality/water_quality_facts_and_stats.pdf>.

37 World Resource Institute (WRI), 'Interactive map of eutrophication and hypoxia', 2013, <www.wri.org/our-work/project/eutrophication-and-hypoxia/interactive-map-eutrophication-hypoxia>.
38 L. Rios, C. Moore and P.R. Jones, 'Persistent organic pollutants carried by synthetic polymers in the ocean environment', *Marine Pollution Bulletin*, 54(8), 2007, pp. 1230–7, <www.sciencedirect.com/science/article/pii/S0025326X07001324>.
39 Australian Institute of Marine Science (AIMS), 'The Great Barrier Reef has lost half of its coral in the last 27 years', June 2012, <www.aims.gov.au/latest-news/-/asset_publisher/MlU7/content/2-october-2012-the-great-barrier-reef-has-lost-half-of-its-coral-in-the-last-27-years>.
40 'Coral Crisis', *Catalyst*, ABC TV, 2006, <www.abc.net.au/catalyst/stories/s1698724.htm>.
41 AIMS, 'Pesticides compound climate risk to reef', 2011, <www.aims.gov.au/docs/research/water-quality/runoff/pesticides-climate-risk.html>.
42 E.N. Veron, *A Reef in Time*, Belknap, Harvard, 2008.
43 UNEP, *Global Chemicals Outlook*.
44 J. Rockstrom et al., 'A safe operating space for humanity', *Nature*, 461, 2009, <www.stockholmresilience.org/download/18.1fe8f33123572b59ab800012568/pb_longversion_170909.pdf>.

Chapter 3

1 R. Roy and V. Agarwal, 'Tainted school lunch kills at least 22 Indian children', *Wall Street Journal*, 18 July 2013, <http://online.wsj.com/article/SB10001424127887323993804578611272207737576.html>.
2 US EPA, 'Pesticide news story: EPA releases report containing latest estimates of pesticide use in the United States', 2011, <http://epa.gov/oppfead1/cb/csb_page/updates/2011/sales-usage06-07.html>.
3 K.S. Schafer, M. Reeves, S. Spitzer and S.E. Kegley, 'Chemical trespass: Pesticides in our bodies and corporate accountability', Pesticide Action Network North America, 2004; 'Potentially harmful pesticides found in all human subjects tested', *Science Daily*, 6 January 2008, <www.sciencedaily.com/releases/2008/01/080104102807.htm>.
4 Centers for Disease Control (CDC), *Fourth National Report on Human Exposure to Environmental Chemicals*, 2009, updated 2013, <www.cdc.gov/exposurereport>.
5 WHO, 'Assessment of capacity in WHO EURO member states to address health-related aspects of chemical safety', 2012, <www.euro.who.int/en/health-topics/environment-and-health/health-impact-assessment/publications/2012/assessment-of-capacity-in-who-euro-member-states-to-address-health-related-aspects-of-chemical-safety>.
6 WHO, 'Dioxins and their effects on human health', May 2010, <www.who.int/mediacentre/factsheets/fs225/en>.

7 EWG, 'Toxic chemicals found in minority cord blood', 2009, <www.ewg.org/news/news-releases/2009/12/02/toxic-chemicals-found-minority-cord-blood>. Full report: <http://static.ewg.org/reports/2009/minority_cord_blood/2009-Minority-Cord-Blood-Report.pdf>.
8 Ibid.
9 S. Nakayama, 'Environmental contaminants and children's health: International collaboration in large scale birth cohort studies', paper presented at CleanUp, 2013, Melbourne.
10 R.H. Weldon, D.B. Barr, C. Trujillo, A. Bradman, N. Holland and B. Eskenazi, 'A pilot study of pesticides and PCBs in the breast milk of women residing in urban and agricultural communities of California', *Journal of Environmental Monitoring*, 13(11), 2011, pp. 3136–44.
11 CRC CARE, 'Save our children from computer toxics: Scientist', media release, 2013, <www.crccare.com/news/save-our-children-from-computer-toxics-scientist>.
12 European Environment and Health Information Service (ENHIS), *Persistent Organic Pollutants in Human Milk*, fact sheet 4.3, 2009, <www.euro.who.int/__data/assets/pdf_file/0003/97032/4.3.-Persistant-Organic-Pollutantsm-EDITED_layouted_V2.pdf>.
13 Australian Department of Sustainability and the Environment, 'Total volatile organic compounds (total VOCs)', 2012, <www.environment.gov.au/atmosphere/airquality/publications/sok/vocs.html>.
14 US EPA, *Indoor Air Toxics*, fact sheet, 2011, <www.epa.gov/ttn/atw/urban/fs_indoor.pdf>.
15 EWG, 'We need safe cosmetics reform now', 2013, <www.ewg.org/enviroblog/2013/05/we-need-safe-cosmetics-reform-now>.
16 University of Toronto, *Environmental Health and Safety*, 2013, <www.ehs.utoronto.ca/resources/HSGuide/Scent.htm#Purpose>.
17 EWG, 'Myths on cosmetics safety', 2013, <www.ewg.org/skindeep/myths-on-cosmetics-safety>.
18 EWG, 'Browsing 17 ingredients', 2014, Skin Deep Cosmetics database, <www.ewg.org/skindeep/>.
19 Cancer Prevention Coalition, 'Cosmetics and personal care products can be cancer risks', 2013, <www.preventcancer.com/consumers/cosmetics/cosmetics_personal_care.htm>.
20 'Taking the "real food" challenge', *RN First Bite*, Radio National, <www.abc.net.au/radionational/programs/rnfirstbite/taking-the-27real-food27-challenge/5041294>.
21 'Real food defined (the rules)', *100 Days of Real Food*, <www.100daysofrealfood.com/real-food-defined-a-k-a-the-rules>.
22 WHO, 'Chemical risks in food', 2012, <www.who.int/foodsafety/chem/en>.
23 UCS, *The Healthy Farm*, April 2013, <www.ucsusa.org/food_and_agriculture/solutions/advance-sustainable-agriculture/healthy-farm-vision.html>.
24 S. Kobylewski and M.F. Jacobson, *Food Dyes: A rainbow of risks*, Center for Science in the Public Interest, 2010, <www.cspinet.org>.

25 EWG, '12 hormone-altering chemicals and how to avoid them', 2013, <www.ewg.org/research/dirty-dozen-list-endocrine-disruptors?inlist=Y&utm_source=201208endocrinedd&utm_medium=email&utm_content=first-link&utm_campaign=toxics>.

26 EWG, 'Finding healthier food', 2013, <www.ewg.org/foodnews/methodology.php>.

27 EU, *Action on Pesticides*, fact sheet, March 2009, <http://ec.europa.eu/food/plant/plant_protection_products/eu_policy/docs/factsheet_pesticides_en.pdf>.

28 European Food Safety Authority (EFSA), 'The 2010 European Union report on pesticide residues in food', *EFSA Journal*, 11(3), 2013, <www.efsa.europa.eu/en/efsajournal/pub/3130.htm>.

29 CDC, 'Antibiotic resistance threats in the United States', 2013, <www.cdc.gov/drugresistance/threat-report-2013/>.

30 European Commission Directorate-General for Health and Consumers, 'Antimicrobial Resistance Policy', 2012, <http://ec.europa.eu/health/antimicrobial_resistance/policy/index_en.htm>.

31 Food Intolerance Network, 2013, <http://fedup.com.au>.

32 Royal Prince Alfred Hospital, *RPAH Elimination Diet Handbook with Food and Shopping Guide*, 2013, <www.sswahs.nsw.gov.au/rpa/allergy/resources/foodintol/handbook.html>.

33 S. Dengate, personal comment, October 2013.

34 US EPA, 'Drinking water contaminants', 2009, <http://water.epa.gov/drink/contaminants/index.cfm#List>.

35 J. Hays, 'Water pollution in China', 2012, <http://factsanddetails.com/china.php?itemid=390>.

36 Ibid.

37 S. Sharma, 'Contaminated water stunting growth of Indian kids: UNICEF', *Hindustan Times*, 14 February 2013, <www.hindustantimes.com/India-news/NewDelhi/Contaminated-water-stunting-children/Article1-1011753.aspx>.

38 European Environment Agency (EEA), 'More than half of EU surface waters below "good" ecological status', November 2012, <www.eea.europa.eu/pressroom/newsreleases/eea-reviews-new-findings-from/highlights/more-than-half-of-eu>.

39 M. Nieuwenhuisen et al., 'Chlorination disinfection byproducts in water and their association with adverse reproductive outcomes: A review', *Occupational Environmental Medicine*, 57(2), 2000, pp. 73–85, <www.ncbi.nlm.nih.gov/pmc/articles/PMC1739910>.

40 'Chemical in soft drinks "can wreck your child's DNA"', *Daily Mail* (UK), <www.dailymail.co.uk/health/article-458011/chemical-soft-drinks-wreck-childs-DNA.html>.

41 S. Labi, 'What chemicals are in your coffee?' *Body+Soul*, 2012, <www.bodyandsoul.com.au/health+healing/news+features/what+chemicals+are+in+your+coffeer,9861>.

42 US National Pesticide Information Center (NPIC), *Older Adults and Pesticides*, fact sheet, 2011, <http://npic.orst.edu/factsheets/olderadults.pdf>.

43 B. Dent, *The Hydrogeological Context of Cemetery Operations and Planning in Australia*, University of Technology Sydney, Sydney, 2002.
44 UNEP, *Global Chemicals Outlook*; note that more than 2 million of these deaths are caused by smoke from indoor cooking fires, outdoor air pollution and tobacco smoke.
45 WHO estimated global malaria deaths at 660,000 in 2010.
46 Blacksmith Institute, 'The world's worst 2013: The top ten toxic threats 2013', 2012, <www.worstpolluted.org/docs/TopTenThreats2013.pdf>.

Chapter 4

1 Ruthann A. Rudel et al., 'Food packaging and Bisphenol A and Bis(2-Ethyhexyl) Phthalate exposure: Findings from a dietary intervention', *Environmental Health Perspectives*, 119(7), 2011, pp. 914–20.
2 S. Frienkel, 'If the food's in plastic, what's in the food?', *Washington Post*, 17 April 2012, <www.washingtonpost.com/national/health-science/trace-chemicals-in-everydayfood-packaging-cause-worry-over-cumulativethreat/2012/04/16/gIQAUILvMT_story.html>.
3 EFSA, 'The 2010 European Union report on pesticide residues in food'.
4 'Food packaging chemicals may be harmful to human health over long term', *Journal of Epidemiology and Community Health*, media release, 20 February 2014.
5 Jane Muncke, John Peterson Myers, Martin Scheringer and Miquel Porta, 'Food packaging and migration of food contact materials: Will epidemiologists rise to the neotoxic challenge?' *Journal of Epidemiology and Community Health*, 20 February 2014.
6 UNEP, *Global Chemicals Outlook*.
7 Sofie S. Christiansen et al., 'Synergistic disruption of external male sex organ development by a mixture of four antiandrogens', *Environmental Health Perspectives*, 117(12), 2009, pp. 1839–46.
8 M. Manikkam et al., 'Plastics derived endocrine disruptors (BPA, DEHP and DBP) induce epigenetic transgenerational inheritance of obesity, reproductive disease and sperm epimutations', *PLoS ONE* 8(1), 2013, <www.plosone.org/article/info%3Adoi%2F10.1371%2Fjournal.pone.0055387>.
9 A. Kortencamp et al., *State of the Art Report on Mixture Toxicity*, EU, The Hague, 2009, <http://ec.europa.eu/environment/chemicals/pdf/report_Mixture%20toxicity.pdf>.
10 D.O. Carpenter, K. Arcaro and D.C. Spink, 'Understanding the human health effects of chemical mixtures', *Environmental Health Perspectives*, 100, 2002, pp. 259–69.
11 SCHER, SCCS, SCENIHR, 'Opinion on the toxicity and assessment of chemical mixtures', 2012, <http://ec.europa.eu/health/scientific_committees/environmental_risks/docs/scher_o_155.pdf>.
12 National Computational Infrastructure, 'Forseeing the unforeseeable', 2012, <http://nci.org.au/researches/forseeing-the-unforseeable>.

13 A.D. Lopez, C.D. Mathers, M. Ezzati, D.T. Jamison and C.J.L. Murray (eds), *Global Burden of Disease and Risk Factors*, WHO, Geneva, 1996.
14 K.R. Smith, C.F. Corvalan and T. Kjellstrom, 'How much global ill health is attributable to environmental factors?' *Epidemiology*, 10, 1999, pp. 573–84.
15 Carpenter, Arcaro and Spink, 'Understanding the human health effects of chemical mixtures'.

Chapter 5

1 M.H. Wong, personal comment, 2013.
2 T. Johnson, 'E-waste dump of the world', *Seattle Times*, 9 April 2006.
3 Greenpeace International, 'Toxic tea party', 23 July 2007, <www.greenpeace.org/international/en/news/features/e-waste-china-toxic-pollution-230707>.
4 C. Carroll, 'High-tech trash', *National Geographic*, January 2008, <http://ngm.nationalgeographic.com/2008/01/high-tech-trash/carroll-text>.
5 M.H. Wong, Y.B. Man, P. Kiddee and R. Naidu, *Adverse Environmental and Health Impacts of Uncontrolled Recycling and Disposal of Electronic Waste*, CRC CARE, Salisbury, SA, 2013.
6 P. Kiddee, R. Naidu and M.H. Wong, 'Electronic waste management approaches: An overview', *Waste Management*, 33, 2013, pp. 1237–50.
7 S. Sthiannopkao and M.H. Wong, 'Handling e-waste in developed and developing countries: Initiatives, practices, and consequences', *Science of the Total Environment*, 463–4, 2012, pp. 1147–53.
8 J.K. Chan and M.H. Wong, 'A review of environmental fate, body burdens, and human health risk assessment of PCDD/Fs at two typical electronic waste recycling sites in China', *Science of the Total Environment*, 463–4, 2012, pp. 1111–23.
9 For example, F. Yang et al., 'Comparisons of IL-8, ROS and p53 responses in human lung epithelial cells exposed to two extracts of PM2.5 collected from an e-waste recycling area, China', *Environmental Research Letters*, 17 May 2011.
10 UNEP, Recycling—From E-waste to Resources, 2009, <www.unep.org/pdf/Recycling_From_e-waste_to_resources.pdf>.
11 WHO, 'Asbestos', 2013, <www.who.int/ipcs/assessment/public_health/asbestos/en/index.html>.
12 C. Buzea, I. Pacheco Blandino and K. Robbie, 'Nanomaterials and nanoparticles: Sources and toxicity', *Biointerphases*, 2(4), 2007, pp. MR17–172.
13 *nanoEHS virtual journal*, 2014, <http://cohesion.rice.edu/centersandinst/icon/virtualjournal.cfm>.
14 S.R. Carpenter, D. Ludwig, and W.A. Brock, 'Management of eutrophication for lakes subject to potentially irreversible change', *Ecological Applications*, 9, 1999, pp. 751–71, <http://dx.doi.org/10.1890/1051-0761(1999)009[0751:MOEFLS]2.0.CO;2>; R.J. Diaz and R. Rosenberg, 'Spreading dead zones and consequences for marine ecosystems', *Science*, 321(5891), 2008, pp. 926–9.

15 C. Boettiger et al., 'Early warning signals: The charted and uncharted territories', *Theoretical Ecology*, 6(3), 2013, pp. 255–64, <http://link.springer.com/article/10.1007%2Fs12080-013-0192-6>.

16 J. Lash, 'Agriculture and "dead zones"', 2007, <www.wri.org/publication/content/7780>.

17 'Dead zone', encyclopaedia entry, *National Geographic Education*, <http://education.nationalgeographic.com.au/education/encyclopedia/dead-zone/?ar_a=1>.

18 See, for example, P.D. Ward, *Under a Green Sky*, Smithsonian Books/Collins, New York, 2007.

19 US EPA, '2006–2007 pesticide market estimates: Usage', <www.epa.gov/pesticides/pestsales/07pestsales/usage2007.htm>.

20 See, for example, W. Zhang, F. Jiang and J. Ou, 'Global pesticide consumption and pollution: With China as a focus', *Proceedings of the International Academy of Ecology and Environmental Sciences*, 1(2), 2011, pp. 125–44.

21 M. Beketov, B. Kefford, R. Schäfer and M. Liess, 'Pesticides reduce regional biodiversity of stream invertebrates', *PNAS*, 2013, <www.pnas.org/cgi/doi/10.1073/pnas.1305618110>.

22 D. Goulson, 'An overview of the environmental risks posed by neonicotinoid insecticides', *Journal of Applied Ecology*, 50(4), 2013, pp. 977–87, <http://onlinelibrary.wiley.com/doi/10.1111/1365-2664.12111/abstract>.

23 S. Oosthoek, 'Pesticides spark broad biodiversity loss: Agricultural chemicals affect invertebrates in streams and soil, even at "safe" levels', *Nature News*, 17 June 2013, <www.nature.com/news/pesticides-spark-broad-biodiversity-loss-1.13214>.

24 C.H. Krupke, G.J. Hunt, B.D. Eitzer, G. Andino and K. Given, 'Multiple routes of pesticide exposure for honey bees living near agricultural fields', *PLoS ONE,* 7(1), 2012, <www.plosone.org/article/info%3Adoi%2F10.1371%2Fjournal.pone.0029268>.

25 R. Van Noorden, 'Europe to ban pesticides in effort to protect bees', *NatureInsight*, 29 April 2013, <http://blogs.nature.com/news/2013/04/europe-to-ban-pesticides-in-effort-to-protect-bees.html>.

26 S. Goldenberg, 'US rejects EU claim of insecticide as prime reason for bee colony collapse', *The Guardian*, 3 May 2013, <www.guardian.co.uk/environment/2013/may/02/us-bee-report-pesticide-eu>.

27 See, for example, the spectrum of opinion from leading UK researchers, Science Media Centre, 'Expert reaction to EU vote on neonicotinoids', <www.sciencemediacentre.org/xpert-reaction-to-eu-vote-on-neonicotinoids>.

28 G. Flores, 'A political battle over pesticides', *The Scientist,* 10 April 2013, <www.the-scientist.com/?articles.view/articleNo/35058/title/A-Political-Battle-Over-Pesticides>.

29 UNEP, *Global Chemicals Outlook*.

30 OPCW, 'Demilitarisation'.

31 J. Bull, 'The deadliness below', *Daily Press*, 30 October 2005.

32 D.M. Bearden, 'U.S. Disposal of Chemical Weapons in the Ocean: Background and issues for Congress', *CRS*, January 2007.

33 MBARI, 'Dangerous unknowns', 2008, <www.mbari.org/news/homepage/2008/chemweapons.html>.

34 UNODC, *World Drug Report 2013*, 2013, <www.unodc.org/unodc/secured/wdr/wdr2013/World_Drug_Report_2013.pdf>.

35 R.A. Nelson et al., 'Alcohol, tobacco and recreational drug use and the risk of non-Hodgkin's lymphoma', *British Journal of Cancer*, 76, 1997, pp. 1532–7.

36 D. Nutt et al., 'Development of a rational scale to assess the harm of drugs of potential misuse', *The Lancet*, 369(9566), 2007, pp. 1047–53.

Chapter 6

1 M. Taylor, 'Lead poisoning of Port Pirie children: A long history of looking the other way', *The Conversation*, 19 July 2012, <http://theconversation.com/lead-poisoning-of-port-pirie-children-a-long-history-of-looking-the-other-way-8296>.

2 P.A. Baghurst et al., 'Environmental exposure to lead and children's intelligence at the age of seven years: The Port Pirie cohort study', *New England Journal of Medicine*, 327, 1992, pp. 1279–84.

3 See, for example, J. Schwarz, 'Low-level lead exposure and children's IQ', *Environmental Research*, 65, 1994, pp. 42–55, <www.rachel.org/files/document/Low-Level_Lead_Exposure_and_Childrens_IQ_A_Met.pdf>; and R. Nevin, 'How lead exposure relates to temporal changes in IQ, violent crime, and unwed pregnancy', *Environmental Research*, 83(1), 2000, <www.sciencedirect.com/science/article/pii/S0013935199940458>.

4 D.C. Bellinger, 'A strategy for comparing the contributions of environmental chemical and other risk factors to children's neurodevelopment', 2012, Environmental Health Perspectives, <http://ehp.niehs.nih.gov/wp-content/uploads/120/4/ehp.1104170.pdf>.

5 A. Chen and W. Hessler, 'Chemical exposures cause child IQ losses that rival major diseases', *Environmental Health News*, 24 February 2012, <www.environmentalhealthnews.org/ehs/newscience/2012/01/2012-0223-chemicals-iq-loss-similar-to-disease>.

6 P. Grandjean and P.J. Landrigan, 'Developmental neurotoxicity of industrial chemicals: A silent pandemic', *The Lancet*, 368(9353), 2006, pp. 2167–78.

7 K. Northstone, C. Joinson, P. Emmett, A. Ness and T. Paus, 'Are dietary patterns in childhood associated with IQ at 8 years of age? A population-based cohort study', *Journal of Epidemiology & Community Health*, 66(7), 2011, pp. 624–8.

8 'Experts call for global overhaul of industrial chemical regulations to ensure children are protected from "silent epidemic" of brain disorders', *Lancet Neurology*, media release, 15 February 2014.

9 Philippe Grandjean and Philip J. Landrigan, 'Neurobehavioural effects of developmental toxicity', *Lancet Neurology*, 15 February 2014.

10 Australian Science Media Centre, 'Chemicals need better regulation to prevent "epidemic" of brain disorders', *Lancet Neurology*, experts respond, 15 February 2014.

11 See, for example, G. Jones and W.J. Schneider, 'Intelligence, human capital, and economic growth: A Bayesian Averaging of Classical Estimates (BACE) approach', *Journal of Economic Growth*, 11(1), 2006, pp. 71–93; G. Jones, 'National IQ and national productivity: The hive mind across Asia', *Asian Development Review*, June 2011, <http://mason.gmu.edu/~gjonesb/JonesADR>.
12 R.J. Herrnstein, C. Murray and F.T. Cullen, *Does IQ Significantly Contribute to Crime?, Taking sides*, Dushkin/McGraw Hill, New York, 1998.
13 K. Chatham-Stephens et al., 'Burden of disease from toxic waste sites in India, Indonesia, and the Philippines in 2010', *Environmental Health Perspectives*, May 2013, <http://ehp.niehs.nih.gov/1206127>.
14 S. Leahy, 'Toxic waste on a par with malaria as a global killer', Inter Press Service, May 2013, <www.ipsnews.net/2013/05/toxic-waste-on-par-with-malaria-as-a-global-killer>.
15 K.S. Schafer and E.C. Marquez, *A Generation in Jeopardy: How pesticides are undermining our children's health and intelligence*, Pesticide Action Network North America, October 2012, <www.panna.org/publication/generation-in-jeopardy>.
16 C. Boyle et al., 'Trends in the prevalence of developmental disabilities in US children, 1997–2008', *Pediatrics*, 127(6), 2011, pp. 1034–42, <www.pediatricsdigest.mobi/content/127/6/1034.full#aff-1>.
17 P.J. Landrigan and J. Forman, 'Chemicals and children's health: The early and delayed consequences of early exposures', paper presented at WHO Forum, Budapest, 2006.
18 'Link between household chemicals and childhood disease?', *Breakfast*, ABC Radio National, 5 August 2011, <www.abc.net.au/radionational/programs/breakfast/link-between-household-chemicals-and-childhood/2928634 August 5>.
19 M. Bouchard et al., 'Attention-Deficit/Hyperactivity Disorder and urinary metabolites of organophosphate pesticides', *Pediatrics*, 125, 2010, <www.pediatricsdigest.mobi/content/125/6/e1270.full?sid=68f0cd91-b8e4-4c25-817c-1d1495e19de9>.
20 WHO, *Youth Violence*, fact sheet 356, August 2011, <www.who.int/mediacentre/factsheets/fs356/en>.
21 M. Elsabbagh et al., 'Global prevalence of autism and other pervasive developmental disorders', *Autism Research*, April 2012, <http://onlinelibrary.wiley.com/doi/10.1002/aur.239/pdf>.
22 CDC, 'Autism spectrum disorders: Data and statistics', 2014, <www.cdc.gov/NCBDDD/autism/data.html>.
23 A. Park, 'Autism rises: More children than ever have autism, but is the increase real?', *Time*, 29 March 2012, <http://healthland.time.com/2012/03/29/autism-rises-more-u-s-children-than-ever-have-autism-is-the-increase-real/#ixzz2XUV1DE7O>.
24 Reuters, 'Traffic pollution tied to autism risk: Study', 2012, <www.reuters.com/article/2012/11/26/us-traffic-pollution-autism-idUSBRE8AP16020121126>; H. Volk et al., 'Residential proximity to freeways and autism in the CHARGE study', *Environmental Health Perspectives*, 119(6), 2011, pp. 873–7, <www.ncbi.nlm.nih.gov/pmc/articles/PMC3114825>.

25 Z.G. Liu, 'A study on the emissions of chemical species from heavy-duty diesel engines and the effects of modern aftertreatment technology', SAE International, 2009, <https://s3.amazonaws.com/drb_website_storage/devinberg.com/liu_09_study.pdf>.

26 M. Weisskopf et al., 'Perinatal air pollution exposure and autism, with new results in the Nurses' Health Study', International Society for Autism Research, May 2013, <https://imfar.confex.com/imfar/2013/webprogram/Paper14885.html>.

27 American Cancer Society (ACS), *Global Cancer Facts and Figures 2011*, 2011, <www.cancer.org/acs/groups/content/@epidemiologysurveilance/documents/document/acspc-027766.pdf>.

28 ACS, *Global Cancer Facts and Figures 2011*, 2011, <www.cancer.org/research/cancerfactsfigures/globalcancerfactsfigures/global-facts-figures-2nd-ed>.

29 N.A. Van Larebeke et al., 'Unrecognized or potential risk factors for childhood cancer', *International Journal of Occupational Environment Health*, 11(2), 2005, pp. 199–201.

30 WHO, 'Asthma', May 2011, <www.who.int/mediacentre/factsheets/fs307/en/index.html>.

31 V. Greenwood, 'Why are asthma rates soaring?', *Scientific American*, 14 April 2011.

32 L.J. Kirmayer et al., 'Explaining medically unexplained symptoms', *Canadian Journal of Psychiatry*, 49(10), 2004, pp. 663–72.

33 Mayo Clinic, 'Morgellons disease: Managing a mysterious skin condition', <www.mayoclinic.com/health/morgellons-disease/SN00043>.

34 A. Wang et al., 'Parkinson's disease risk from ambient exposure to pesticides', *European Journal of Epidemiology*, 26(7), 2011, pp. 547–55.

35 K. Hayden et al., 'Occupational exposure to pesticides increases the risk of incident AD: The Cache County study', *Neurology*, 74(19), 2010, pp. 1524–30.

36 WHO, 'Dementia: A public health priority', WHO and Alzheimer's Disease International, 2012, <www.who.int/mental_health/publications/dementia_report_2012/en>.

37 'Study suggests possible association between cardiovascular disease, chemical exposure', media release, JamaNet, 3 September 2012, <http://media.jamanetwork.com/news-item/study-suggests-possible-association-between-cardiovascular-disease-chemical-exposure>.

38 WHO, *Principles and Methods for Assessing Autoimmunity Associated with Chemicals*, WHO, Geneva, 2006.

39 EWG, 'Pollution in minority newborns', <www.ewg.org/research/minority-cord-blood-report>.

40 EEA, 'Increase in cancers and fertility problems may be caused by household chemicals and pharmaceuticals', May 2012, <www.eea.europa.eu/pressroom/newsreleases/eea-reviews-new-findings-from/pressroom/newsreleases/increase-in-cancers-and-fertility>.

41 National Institute of Environmental Health Sciences (NIEHS), 'Endocrine disruptors', 2013, <www.niehs.nih.gov/health/topics/agents/endocrine>.

42 R. McKie, '£30bn bill to purify water system after toxic impact of contraceptive pill', *The Guardian*, 3 June 2012, <www.theguardian.com/environment/2012/jun/02/water-system-toxic-contraceptive-pill>.

NOTES

43 T. Colborn, 'The fossil fuel connection', The Endocrine Disruption Exchange, 2010, <http://endocrinedisruption.org/endocrine-disruption/the-fossil-fuel-connection>.

44 A. Bergman et al., *State of the Science of Endocrine Disrupting Chemicals 2012*, WHO/UNEP, Geneva, 2013.

45 EWG, 'Dirty dozen list of endocrine disruptors: 12 hormone-altering chemicals and how to avoid them', 2013, <www.ewg.org/research/dirty-dozen-list-endocrine-disruptors?inlist=Y&utm_source=201208endocrinedd&utm_medium=email&utm_content=first-link&utm_campaign=toxics>.

46 WHO, *World Cancer Report 2014*, Bernard W. Stewart and Christopher P. Wild (eds), February 2014, <www.iarc.fr/en/publications/books/wcr/index.php>.

47 IARC/WHO, 'Global battle against cancer won't be won with treatment alone: Effective prevention measures urgently needed to prevent cancer crisis', WHO press release 224, 3 February 2014, <www.iarc.fr/en/media-centre/pr/2014/pdfs/pr224_E.pdf>.

48 Cancer Council of NSW, 'Hope: Turning the page on cancer', 2013, <http://hope.cancercouncil.com.au/?gclid=CJ3F8JK14LgCFUhapQodB2YALQ>.

49 ATSDR, *Chemicals, Cancer and You*.

50 US Department of Health and Human Services National Toxicology Program, *12th Report on Carcinogens*, 10 June 2011, <http://ntp.niehs.nih.gov/?objectid=72016262-BDB7-CEBA-FA60E922B18C2540>.

51 Breast Cancer Fund, 'Chemicals and radiation linked to breast cancer', <www.breastcancerfund.org/clear-science/radiation-chemicals-and-breast-cancer>.

52 Interagency Breast Cancer and Environmental Research Coordinating Committee (IBCERCC), 'Breast cancer and the environment: Prioritizing prevention', 2013, <www.niehs.nih.gov/about/boards/ibcercc>.

53 University of California, 'Pathways to Breast Cancer', 2010, <http://coeh.berkeley.edu/greenchemistry/cbcrpdocs/executive_summary.pdf>.

54 'Breast cancer and chemicals', *The Health Report*, ABC Radio National, 1 March 2010, <www.abc.net.au/radionational/programs/healthreport/breast-cancer-and-chemicals/3119582#transcript>.

55 WHO, *Depression*, fact sheet, October 2012, <www.who.int/mediacentre/factsheets/fs369/en>.

56 E. Rivero, 'Nearly 5% of U.S. population suffers from persistent depression or anxiety', media release, UCLA, 1 December 2008, <www.newsroom.ucla.edu/portal/ucla/nearly-5-percent-of-the-u-s-population-72195.aspx>.

57 Harvard Medical School, 'What causes depression?', 2011, <www.health.harvard.edu/newsweek/what-causes-depression.htm>.

58 See, for example, C. Kresser, 'The "chemical imbalance" myth', 2008, <http://chriskresser.com/the-chemical-imbalance-myth>.

59 L. Stallones and C. Beseler, 'Pesticide poisoning and depressive symptoms among farm residents', *Annals of Epidemiology*, 12(6), 2002, pp. 389–94, <www.ncbi.nlm.nih.gov/pubmed/12160597?dopt=Abstract>.

60 T.A. Rehner et al., 'Depression among victims of South Mississippi's methyl parathion disaster', *Health & Social Work*, February 2000, pp. 33–40, <www.naswdc.org/pressroom/events/911/rehner.asp>.

61 M. Porta and D.H. Lee, *Review of the Science Linking Chemical Exposures to the Human Risk of Obesity and Diabetes*, ChemTrust UK, March 2012, <www.wecf.eu/download/2012/March/CHEMTrustObesityDiabetesSummaryReport.pdf>.
62 R.R. Newbold, 'Impact of environmental endocrine disrupting chemicals on the development of obesity', *Hormones*, 9(3), 2010, pp. 206–17.
63 WHO, *Obesity and Overweight*, fact sheet, March 2013, <www.who.int/mediacentre/factsheets/fs311/en>.
64 J.C.F. Siow, *Health Impacts of Persistent Organic Pollutants and/or Heavy Metals*, CRC CARE, 2013.
65 S. Gilbert, 'Epigenetics', *Toxipedia*, 28 February 2013, <http://toxipedia.org/display/toxipedia/Epigenetics>.
66 A. Baccarelli and V. Bollati, 'Epigenetics and environmental chemicals', *Current Opinions in Pediatrics*, 21(2), 2009, pp. 243–51.
67 R. Holliday, 'Epigenetics: A historical overview', *Epigenetics*, 1(2), 2006, <www.braintrusters.com/journals/epigenetics/hollidayEPI1-2.pdf>.
68 'Food addiction and the brain', *The Health Report*, ABC Radio National, <www.abc.net.au/radionational/programs/healthreport/food-addiction/4865260>.
69 L. Hou, X. Zhang, D. Wang and A. Baccarelli, 'Environmental chemical exposures and human epigenetics', *International Journal of Epidemiology*, 41(1), 2012, pp. 79–105, <www.ncbi.nlm.nih.gov/pmc/articles/PMC3304523/#dyr154-B1>.
70 Ibid.
71 Health and Environment Linkages Initiative (HELI), 'Using economic valuation methods for environment and health assessment', WHO, n.d., <www.who.int/heli/economics/valmethods/en/index.html>.
72 UNEP, *Global Chemicals Outlook*, p. 27.
73 C.P. Wild, 'Complementing the genome with an "exposome": The outstanding challenge of environmental exposure measurement in molecular epidemiology', *Cancer, Epidemiology, Biomarkers & Prevention*, 14(8), 2005, pp. 1847–50.
74 P. Grandjean and P.J. Landrigan, 'Developmental neurotoxicity of industrial chemicals: A silent pandemic', *The Lancet*, 368(9553), 2006, pp. 2167–78.

Chapter 7

1 Smith and Smith, *Minamata*.
2 M. Yamaguchi and A. Chang, 'Fukushima tsunami plan: Japan nuclear plant downplayed risk', *Huffington Post*, 29 July 2013, <www.huffingtonpost.com/2011/03/27/fukushima-tsunami-plan-japan_n_841222.html>.
3 Environmental Working Group, 'Cancer-causing chemical found in 89 percent of cities sampled', 20 December 2010, <www.ewg.org/chromium6-in-tap-water>.
4 P. Lagadec, 'From Seveso to Mexico and Bhopal: Learning to cope with crises', 1987, <www.patricklagadec.net/fr/pdf/Mexico_Bhopal.pdf>.

5 'Directive 2012/18/Eu of the European Parliament and of the Council of 4 July 2012 on the control of major-accident hazards involving dangerous substances, amending and subsequently repealing Council Directive 96/82/EC', *Official Journal of the European Union*, <http://eur-lex.europa.eu/LexUriServ/LexUriServ.do?uri=OJ:L:2012:197:0001:0037:EN:PDF>.
6 M.J. Peterson, *Case Study: Bhopal plant disaster*, International Dimensions of Ethics Education Case Study Series, March 2009.
7 International Campaign for Justice in Bhopal, <http://bhopal.net>.
8 Stockholm Convention, 2013, <http://chm.pops.int/Convention/ThePOPs/The12InitialPOPs/tabid/296/Default.aspx> and <http://chm.pops.int/Implementation/NewPOPs/TheNewPOPs/tabid/672/Default.aspx>.
9 R. Ruiz, 'Industrial chemicals lurking in your bloodstream', *Forbes Magazine*, 21 January 2010, <www.forbes.com/2010/01/21/toxic-chemicals-bpa-lifestyle-health-endocrine-disruptors.html>.
10 UNEP, *Global Chemicals Outlook*.
11 Ibid.
12 'Chemical time bomb', *Four Corners*, ABC TV, 22 July 2013.
13 B.M. Decker and W.C. Triplett, 'China's poisonous exports', *Washington Times*, 14 November 2011, <www.washingtontimes.com/news/2011/nov/15/chinas-poisonous-exports>.
14 American Chemistry Council, 'Responsible care guiding principles', 2013, <http://responsiblecare.americanchemistry.com/Responsible-Care-Program-Elements/Guiding-Principles/default.aspx>.
15 See Royal Society of Chemistry, Regulation of the Profession and Code of Conduct, <www.rsc.org/images/code-of-conduct_tcm18-5101.pdf>.
16 Royal Australian Clinical Institute Incorporated, *Code of Ethics*, <www.raci.org.au/document/item/90>.
17 European Chemical Industry Council, 'The chemical industry's commitment to sustainability', <www.cefic.org/Responsible-Care>.
18 'Angry Mermaid Award: European Chemical Industry Council (CEFIC)', <www.angrymermaid.org/CEFIC.html>.
19 See US EPA, 'Sustainable futures', <www.epa.gov/oppt/sf> and 'About EPA', <www.epa.gov/aboutepa/our-mission-and-what-we-do>.
20 UNEP, *Global Chemicals Outlook*.
21 Plan of Implementation of the World Summit on Sustainable Development, 2012 <www.un.org/esa/sustdev/documents/WSSD_POI_PD/English/WSSD_PlanImpl.pdf>.

Chapter 8

1 J. Haynes, *Socio-economic Impact of the Sydney 2000 Olympic Games*, Centre d'Estudis Olímpics UAB, Barcelona, 2001.
2 Sydney Olympic Park Authority, 'Site remediation', 2011, <www.sopa.nsw.gov.au/our_park/history_and_heritage/site_remediation>.

3 J. Hunt et al., *Homebush Bay Sediment Remediation: A case study*, Thiess Services Pty Ltd, Sydney, 2009.
4 R. Naidu, *Contamination: Big risks, bigger opportunities*, CRC CARE, Salisbury, SA, 2012.
5 UNEP, 'Strategic approach to international chemicals management', <www.saicm.org>.
6 Johannesburg Declaration on Sustainable Development, <www.unescap.org/esd/environment/rio20/pages/Download/johannesburgdeclaration.pdf>.
7 UNEP, *Global Chemicals Outlook.*
8 CSIRO, 'Life cycle assessment', 2011, <www.csiro.au/en/Organisation-Structure/Flagships/Sustainable-Agriculture-Flagship/Life-cycle-assessment.aspx>.
9 The Sustainable Scale Project, 'Material flow analysis', 2003, <www.sustainablescale.org/conceptualframework/understandingscale/Measuringscale/Materialflowanalysis.aspx>.
10 UN Framework Convention on Climate Change (UNFCCC), 'Multicriteria analysis (MCA)', 2014, <http://unfccc.int/files/adaptation/methodologies_for/vulnerability_and_adaptation/application/pdf/multicriteria_analysis__mca_pdf.pdf>.
11 Organisation for Economic Cooperation and Development (OECD), 'Environmental policy tools and evaluation: Extended producer responsibility', 2001, <www.oecd.org/env/tools-evaluation/extendedproducerresponsibility.htm>.
12 US EPA, 'Green chemistry', 2014, <www.epa.gov/greenchemistry>.
13 'Manufacturing sustainability', *Green Manufacturing News*, 2014, <www.greenmfgnews.com>.
14 'Green building', *Wikipedia*, <http://en.wikipedia.org/wiki/Green_building>.
15 US EPA, 'Integrated pest management', 2012, <www.epa.gov/pesticides/factsheets/ipm.htm>.
16 International Society for Industrial Ecology, <www.is4ie.org>.
17 Zero Waste Australia, 'What does zero waste mean to your business?', 2014, <http://zerowasteaustralia.org>.
18 L.H. Loong, 'Clean and Green Singapore', launch speech, 2014, <http://app.mewr.gov.sg/web/contents/contents.aspx?contid=1899>.
19 US EPA, 'Cleaning up the nation's hazardous wastes sites', 2014, <www.epa.gov/superfund/index.htm>.
20 US EPA, 'National priorities list (NPL)', 2014, <www.epa.gov/superfund/sites/query/queryhtm/nplfin2.htm>.
21 C.C. Harris, 'The carcinogenicity of anticancer drugs: A hazard in man', *Cancer*, 37(2), 1976, Suppl., pp. 1014–23.
22 R. Moynihan and D. Henry, 'The fight against disease mongering: Generating knowledge for action', *PLoS Med*, 3(4), 2006, <www.ploscollections.org/article/info%3Adoi%2F10.1371%2Fjournal.pmed.0030191;jsessionid=C6610FF39B7EFE9A4539072858BBBAD2>.
23 CDC, 'Antibiotic resistance threats in the United States'.

24 J. Latham, 'Science and social control: Political paralysis and the genetics agenda', *Independent Science News*, 3 August 2013, <www.independentsciencenews.org/science-media/science-and-social-control-political-paralysis-and-the-genetics-agenda>.
25 Chemical Heritage Foundation, 'About women in chemistry', <www.chemheritage.org/discover/online-resources/women-in-chemistry/about.aspx>.
26 Nobel Prize Organisation, 'All Nobel Prizes in Chemistry', 2014, <www.nobelprize.org/nobel_prizes/chemistry/laureates>.
27 Catalyst, *Women in the Sciences*, New York, Catalyst, 2013, <www.catalyst.org/knowledge/women-sciences>.
28 S. Rodnes and M. Lidén, *A Chemical Imbalance*, 'Watch the film', 2013, <http://chemicalimbalance.co.uk/project/watch-the-film>.
29 T. Colborn, 'The fossil fuel connection', The Endocrine Disruption Exchange, 2010, <http://endocrinedisruption.org/endocrine-disruption/the-fossil-fuel-connection>.
30 T. Shirvani, X. Yan, O. Inderwildi, P. Edwards and D. King, 'Life cycle energy and greenhouse gas analysis for algae-derived biodiesel', 2011, <www.smithschool.ox.ac.uk/research/library/TShirvani-EES-Manuscript-2011-web.pdf>.
31 US Department of Energy, 'National Algal Biofuels Technology Roadmap', 2010, <www1.eere.energy.gov/bioenergy/pdfs/algal_biofuels_roadmap.pdf>.
32 A. Green, personal comment, November 2013.
33 A. Green, *Plant Oils: A sunrise industry*, AAOCS, November 2013.

Chapter 9

1 N. and J. Chuda, *Healthy Child Healthy World*, 2014, <http://healthychild.org>.
2 C.R. Sharp et al., 'Parental exposures to pesticides and risk of Wilms' tumor in Brazil', *American Journal of Epidemiology*, 141(3), 1995, pp. 210–17, <www.ncbi.nlm.nih.gov/pubmed/7840094>.
3 L. Mott, F. Vance and J. Curtis, *Handle with Care: Children and environmental carcinogens*, Natural Resource Defense Council, New York, 1994.
4 P. Teilhard de Chardin, *The Phenomenon of Man*, Harper Perennial, New York, 1955.
5 J. Cribb, 'Thinking at species level', paper presented to Fenner Conference on Population and Climate Change, Canberra, Australia, October 2013.
6 Cisco, 'Cisco's Visual Networking Index forecast projects nearly half the world's population will be connected to the internet by 2017', Cisco, 2013, <http://newsroom.cisco.com/release/1197391>.
7 J. Cribb, 'Taxonomy: New name needed for unwise homo?', *Nature*, 476(7360), p. 282, <www.nature.com/nature/journal/v476/n7360/full/476282b.html>.
8 Not to be confused with Clean Up the World, a laudable organisation with 35 million volunteers in 120 countries with a particular focus on removing rubbish and planting trees to improve the environment: see <www.cleanuptheworld.org/en>.

9 N. Oreskes and E. Conway, *Merchants of Doubt*, Bloomsbury, London, 2010.
10 A. Jogalekar, 'Where's the chemistry lobby? On why we need a National Center for Chemical Education', *Scientific American*, July 2013, <http://blogs.scientificamerican.com/the-curious-wavefunction/2013/07/11/wheres-the-chemistry-lobby-on-why-we-need-a-national-center-for-chemical-education>.
11 W. Koch, 'Ten retailers urged to pull potentially toxic products', *USA Today*, 10 April 2013, <www.usatoday.com/story/news/nation/2013/04/09/retailers-products-toxic-chemicals/2067113>.
12 L. Oates, *Organic Food Survey*, Royal Melbourne Institute of Technology, Melbourne, 2010, <www.rmit.edu.au/wellness/organicresearch>.
13 M. Wilsey and S. Lichtig, 'The Nike controversy', <www.stanford.edu/class/e297c/trade_environment/wheeling/hnike.html>.
14 'Fashion victims', *Four Corners*, ABC TV, 24 June 2013, <www.abc.net.au/4corners/stories/2013/06/25/3785918.htm>.
15 'The ethical shopper', *Life Matters*, ABC Radio National, 25 June 2013, <www.abc.net.au/radionational/programs/lifematters/the-ethical-shopper/4777026>.
16 AVAAZ.orgm, <www.avaaz.org/en>.
17 Sum Of Us, <http://sumofus.org/about>.
18 GetUp!, <www.getup.org.au>.
19 Safer Chemicals.org, <www.saferchemicals.org/about/who.html>.
20 Greenpeace, 'Eliminate toxic chemicals', <www.greenpeace.org/international/en/campaigns/toxics>.
21 Friends of the Earth, <www.foei.org/en>.
22 Pesticide Action Network (PAN), 'PAN International', <www.panna.org/our-community/pan-international>.
23 PAN, <www.pesticideinfo.org>.
24 Circle of Blue, 'Choke point: Index dry fields, dirty water', <www.circleofblue.org/waternews>.
25 International River Foundation, <www.riverfoundation.org.au/index.php>.
26 World Wildlife Fund (WWF), <http://worldwildlife.org>.
27 R. Naidu, *Global Contamination Research Initiative (GCRI): A proposal for a new global research initiative addressing one of the most serious human impacts on the planet and on our own future*, CRC CARE/University of South Australia, Salisbury, SA, 2013.
28 UN, *The Universal Declaration of Human Rights*, <www.un.org/en/documents/udhr>.

INDEX

INDEX

INDEX